Preface To Instructors

Why Qualitative *and* Instrumental Analysis?

First-year chemistry students are usually intrigued by the laboratory section of their general chemistry course—often the first hands-on chemistry they have ever experienced. The classical qualitative analysis scheme is probably more successful at this than most sequences or groups of experiments because it holds a number of pleasant and interesting phenomena: colorful changes, solid, liquid, and gas evolution and unknown results. The chemical answers that come by students' hands and totally under students' command give them a feeling of success and control unattainable in many other kinds of experiments that are often used in the freshman laboratory sequence.

The classical Qual scheme is now seldom used in real world chemical analysis; this is a dose of chemical reality that we seldom teach our students. The advent of instrumentation has bypassed the use of wet methods in favor of computer driven and extremely sensitive devices that routinely detect chemical species in parts per billion—well below the detection limits of the Qual scheme. Our growing understanding of the contamination of our planet and the dangers of "trace" levels of toxic materials in our environment dictates that we be as exacting as possible in our environmental analyses. Indeed, our regulations are requiring the detection of smaller and smaller amounts of poisons in our soils, water, and air as we discover the dangerous effects of amounts that were basically *undetectable* a few decades ago.

This concentration by analytical chemists on instrumentation does not mean that the Qual scheme is outdated as a teaching tool for chemistry students; nor does the complexity of this instrumentation mean that chemistry students' introduction to real world environmental analysis need wait until the senior year. On the contrary, the ability of the qualitative analytical scheme, flow chart and all, to present logical processes of analyses and to introduce students to descriptive chemistry is as strong as ever, especially when it is successfully integrated with theoretical concepts like solubility, K_{sp}, and matter's interaction with light. Likewise, even though the most modest analytical instruments (beyond the ubiquitous Spectronic 20® spectrophotometer) are usually not available to freshman lab classes, this does not mean that students can't be conceptually introduced to simple treatments of major analytical instruments. After all, the workings of these instruments are based on concepts that we teach them in the freshman lecture: specifically light emission and absorption, electrostatic attractions and repulsions, conductivity, surface adsorption, and so on. The real world, "big picture" techniques introduced in this book can be taught to freshman either in lab or in lecture even though they might not get their hands on the instrumental knobs for quite a while. This book makes an effort to get students as close as possible to the process of qualitative *and* instrumental analysis in such a way that any mutual exclusivity that we have mistakenly imposed upon these areas will be broken down.

This book uses concepts introduced in the first and second semester of freshman chemistry, and therefore, the in-depth discussions of the theory of qualitative analysis, solubility, equilibrium, or complexation for instance, are left, it is believed appropriately, to the lecture text. It is felt that the supplemental nature of this work demands that the student does not pay for this material twice: once in the lecture text and again in a source of this nature that is meant to augment the lecture material. The concept of common ions, for instance, are only accessed in the chapter problems and not in the text though it is an important part of qualitative analysis and will be covered in depth in freshman lecture courses. Instead, the main stay of each chapter is the **environmental importance** of chemical species and the **instrumental means** of sensitively detecting them. It is felt that it is this emphasis that will interest and engage students in a manner that is both realistic and appropriate given the present role that chemistry plays in our lives and the role it will more strongly play in the future.

The quantity of the material here is in general intended to augment the lecture and laboratory classes with information that is not usually made available to freshman and is, therefore, designed to support that material not supplant it. The book's infrastructure of the classical qualitative scheme will support those sequences that depend on the qualitative laboratory emphasis now or broaden those programs that have excluded the qualitative scheme in the past. The procedures are not meant to replace a complete laboratory manual but instead are intended to enlarge the choice of experiments that are available in a freshman laboratory sequence. Each chapter contains qualitative procedures that are detailed enough to be used as two or three hour stand-alone laboratories in their own right if the introductory and instrumental material is discussed in a recitation, pre-lab or post-laboratory setting and if the unknown group experiments are carried out. However, if two or three of the instrumentation experiments are completed by the students either alone or in groups, this book describes a full stand alone semester of work encompassing qualitative, environmental, and instrumental analysis.

Organization

This book opens with an introductory chapter that details concepts of solvation, concentration, solubility, and solubility product—the heart of qualitative analysis. This is the only "theory" based chapter however, and it ends with a practical discussion of laboratory safety and first aid. Each subsequent chapter begins with an introduction to the environmental importance of a specific toxic metal of each classical Qual group, mercury (Group I), cadmium (Group II), chromium (Group III), etc. and includes step by step procedures of chemical separation of selected members of the group. This section often culminates in the student's construction of a Qual scheme flow chart that may be used to detect an unknown designed and handed out by the instructor.

The second half of each chapter involves the introduction of the major instrumental method used in the environmental analysis of the chapter's subject metal. [In a number of cases these analyses have been derived from procedures that are presently being used at one of the more prominent environmental laboratories in Texas.] Each method's introduction includes the theoretical concepts necessary to understand the

Qualitative and Instrumental Analysis of Environmentally Significant Elements

THOMAS G. CHASTEEN
Sam Houston State University
Huntsville, Texas

JOHN WILEY & SONS, INC.

New York　　Chichester　　Brisbane　　Toronto　　Singapore

Coventry University

Copyright © 1993 by John Wiley & Sons, Inc.

All rights reserved.

Reproduction or translation of any part of this work beyond that permitted by Sections 107 and 108 of the 1976 United States Copyright Act without the permission of the copyright owner is unlawful. Requests for permission or further information should be addressed to the Permissions Department, John Wiley & Sons, Inc.

ISBN 0-471-58649-8

Printed in the United States of America

10 9 8 7 6 5 4 3 2 1

fundamental workings of the instrument used, for example, light absorption or emission, conductivity, or a simple theory of chromatography. This section includes a description of the instrument's major components, complete with a schematic diagram of their interactions in a working instrument. Paths for light, signal, or analyte flow are clearly detailed in order for the student to follow the sequence of events that occur in an analysis—not just a frozen template of stylized (schematic) symbols with little or no meaning.

This part of the chapter also often includes a description of a *quantitative* tool used by chemists in routine instrumental analyses: standards, calibration, spectral display, etc. This will prepare students for the upcoming experimental sequences that they may encountering in sophomore level quantitative analysis.

By request from reviewers, each instrumental section also includes a general description of a simple instrumental experiment involving the technique under study. These experiments are by their nature very generalized because of the differences between instrument setup and procedures that vary greatly from one instrument manufacturer to another. Each procedure will have to be worked out in detail by the instructor before hand if these experiments are to be carried out; however, this section is included in hopes that freshman students can be exposed to at least a few of the real instruments that form the lion's share of the everyday chemist's life.

Each chapter ends with a number of questions about the environmental, procedural, and instrumental concepts that were used in that chapter. More probing questions are designated with an asterisk. References are made in the text to appendices including a table of solubility products and significant figure and rounding rules. The significant figure appendix incorporates exercises with the answers included for the students. An appendix that details some generalized waste minimization and disposal procedures and another that details the reagents used in the text's analyses are included, again due to the manuscript review process.

Finally, a glossary and an index of key terms are included to help students use the book to their best advantage. Terms found in the glossary are marked in boldface in the text.

The idea that all chemists are protective of their environment is obvious if one visits their laboratories—laboratories where toxic materials are handled carefully and safely every day. The application of this common sense approach of protecting our immediate surroundings is slowly being applied to the planet as a whole. It is called the environmental movement. It is hoped that in teaching that there is no real difference between what is inside and outside the laboratory walls will have a profound and important effect on the chemists who use this book.

Thomas G. Chasteen
July 1993

Contents

Chapter 1	Qualitative Separation and Solubility	**1**
Chapter 2	Toxic Mercury and Group I Cations	**11**
Chapter 3	Environmental Cadmium	**27**
Chapter 4	Chromium: Toxic or Essential	**43**
Chapter 5	Barium and Group IV	**61**
Chapter 6	Group V: Acid Rain and the Ion Pump	**79**
Chapter 7	Qualitative Analysis of Anions	**95**
Appendices		
1	Solubility Product Values at 25° C	**105**
2	Significant Figures	**107**
3	Chemical Reagents	**115**
4	Toxic Waste Disposal	**121**
Glossary		**125**
Index		**129**

Chapter 1

Qualitative Separation and Solubility

INTRODUCTION

The separation of molecules can take place in many different ways. Organic liquids can be separated from a mixture by distillation using the differences in boiling points to separate the constituents: Careful heating of a solution containing two liquids will allow you to boil away (and therefore collect and partially identify) one compound and leave the other behind. As we will see soon, this is impossible for dissolved ionic compounds which must be pursued in a different way.

Separating mixtures of gases like those in the atmosphere is a bit more complicated but, again, can be accomplished by using temperature as the tool to cause gases with higher boiling points to selectively condense and allow lower boilers to remain as gases. In this way, oxygen can be separated from nitrogen in the atmosphere where they are both present in great abundance (O_2 ~21 percent and N_2 ~78 percent). This means that pure liquid nitrogen, which has a variety of uses as a liquid coolant, can be easily collected straight from the air, and thus makes it almost as cheap as milk in price.

Separating ions (or ionic complexes) dissolved in water actually presents the easiest problem of these three. Using straight forward, well-known chemical reactions that control the amount of a dissolved solid that can exist in water solution, molecules can be made to selectively fall out of the solution (precipitate) leaving behind other molecules that are still dissolved in solution. Which compounds precipitate and which remain in solution as well as their colors and consistencies provides a simple means of identification, a means of **qualitative analysis**.

WATER AS A SOLVENT

First let's discuss why some compounds dissolve in water and some don't. The water molecule has a specific structural characteristic that makes it an excellent **solvent** ("the dissolver"). Water, H_2O, is a **polar** molecule. This means that the physical arrangements of the atoms in the H_2O molecule give it a bent, nonsymmetrical shape. Therefore, even though the water molecule's overall structure is neutral (uncharged—it is not an ion) it has a relative negative end (the oxygen atom and its electrons) and a relative positive end (the hydrogen atoms). This molecule, including its atoms, their electrons, and the positions of the bonds is therefore polar: the molecule has a net **dipole** (or dipole moment). This polarity is signified in the adjacent drawing by the δ^+ or δ^- notation at either end of the water dipole.

The polarity of water has significant implications in qualitative analysis of aqueous systems. Using the old maxim that like dissolves like, polar or charged species (compounds, molecules, or ions) will interact with water in a way different from nonpolar or neutral species. This is true because there is an electrostatic attraction (attraction between charges) between water (the solvent) and a compound dissolved in it (the **solute**). For species with a net charge, **ions**, this attraction occurs between the oppositely charged δ negative end of water and the positive charge on the ion (called a **cation**). For **anions** (with a negative charge) the attraction is between the δ positive end of water and the negative charge of the anion. This interaction is directly reflected in the **solubility** of these compounds in water. Water, in effect, surrounds the solute molecules and *dis*solves it because of its attraction for the solute's structure. Here is a possible arrangement of water around ions. A cation is represented by a plus sign and an anion by a minus sign.

Remember, too, that this is a two-dimensional representation of a three-dimensional situation. Each of these ions also has two additional water molecules associated with it: one behind the page and one in front of the page. It may be easier to think of this situation as that of a three-dimensional cube, with the ion in the center of the cube and a water molecule on each of the six faces pointing oxygens or hydrogens at the ion depending on whether it's a cation or anion. The **polarity** of water also allows it to dissolve uncharged molecules that are themselves polar. Molecules which are polar like water, for instance ethanol (C_2H_5OH) or formaldehyde (CH_2O) are very soluble in water. Using the same logic, molecules which are neither charged nor polar are, in general, not water soluble. A good example is carbon tetrachloride (CCl_4). Carbon tertrachloride is considered to be insoluble or immiscible with water. Notice that an effort has been made to try to represent this three dimensional (nonpolar) molecule on the two-dimensional page more clearly using a dark colored wedge to represent the bond to the chlorine atom in front of the page and a light colored wedge to represent the bond to the chlorine atom behind the page. Again, it is important to remember that these molecules are three-dimensional entities.

This then sets the stage for a chemist to be able to determine, using the like dissolves like rule of thumb, which molecules should be water soluble and which probably are not. However, the heart of qualitative analysis is not simply whether or not a compound is soluble; qualitative analysis focuses on how soluble different compounds are in relation to each other.

The fact that some molecules or ionic compounds do dissolve in water does not mean that they are infinitely soluble: solutions do in fact become **saturated**. At this point no more solute will dissolve no matter how much is added to the solution. Any added solid will simply fall to the bottom of the container. In fact, as you will soon learn, the limited solubility of different substances can be used to the best advantage by a chemist who wants to separate compounds based upon their individual or differential solubilities.

CONCENTRATION OF A SOLUTION

The basic unit of concentration used in this book is **molarity**, M, the number of moles of solute per liter of solution. The solutes will be various acids (like HCl—hydrochloric acid), bases (like NaOH—sodium hydroxide), and salts (like $AgNO_3$). Calculate the molarity of a solution of 23.45 grams of NaOH dissolved in one liter of water (i.e. enough water to equal to a final volume of one liter):

$$\frac{23.45 \text{ grams NaOH}}{39.91 \text{ g/mole}} = \frac{0.5876 \text{ moles NaOH}}{1 \text{ liter volume}} = 0.5876 \, M \text{ sodium hydroxide}$$

SOLUBILITY PRODUCT AND EQUILIBRIUM

Molar solubility is the number of moles of a compound that will dissolve in one liter of water—the molarity of the saturated solution of a particular compound. Like all solubility this depends on temperature. For a compound that has a one-to-one **stoichiometric** relationship with its dissolved ions, the molar solubility of the compound is equal to the molarity of any of its dissolved ions in a saturated solution. You will see how this is calculated below; however, the relationship that is usually focused on for solutions of this kind is called the solubility product. The solubility product for a particular substance can be determined by analyzing the ions in a saturated liquid in contact with a solid.

When a slightly soluble solid, for instance, iron(II) sulfide is added to pure water it begins to dissolve. This dissolution yields, in the case of iron(II) sulfide, dissolved (that is, solvated and separated) iron ions and sulfide ions:

$$FeS \leftrightarrow Fe^{2+} + S^{2-}$$

Notice that the water molecules surrounding each of the ions are routinely left out of the balanced equation. This dissolution process, as noted above, does not go on indefinitely however; the concentration of iron and sulfide ions increase until the solution becomes saturated. At this point an **equilibrium** is established. The rate at which iron sulfide is dissolving equals the rate at which iron and sulfide ions are coming together to reform the solid. This process is going on in both directions at once, which is why this is often called dynamic equilibrium.

The equilibrium relationship is denoted in the balanced chemical reaction by the use of a double headed arrow. This means that at equilibrium, the concentrations of iron and sulfide ions in the solution (in contact with solid FeS) are constant. The value that expresses this relationship is called the **solubility product constant**: the mathematical product of the concentration of iron and sulfide ions dissolved in solution at a particular temperature. For iron(II) sulfide this is calculated like this:

$$K_{sp} = [Fe^{2+}][S^{2-}]$$

The square brackets mean concentration in units of molarity, and no operator between the brackets means multiplication. *Notice that the "concentration" of the solid, FeS, is excluded from this relationship.* Using the value of K_{sp} from Appendix 1 and y to represent unknown values, the concentration of iron (and sulfide ion) in a saturated solution can be calculated:

$$K_{sp} = 6.00 \times 10^{-18} = [Fe^{2+}][S^{2-}] = (y)(y)$$

Solving for y: $\qquad 6.00 \times 10^{-18} = y^2$ and $y = 2.45 \times 10^{-9}$

$$K_{sp} = 6.00 \times 10^{-18} = [Fe^{2+}][S^{2-}] = (2.45 \times 10^{-9})(2.45 \times 10^{-9})$$

The concentration of iron dissolved in a saturated solution of iron sulfide is therefore 2.45×10^{-9} M. (Review rounding and significant figure rules in Appendix 2 if necessary.) Since y also represents sulfide anion, $[S^{2-}] = 2.45 \times 10^{-9}$ M. And furthermore, since the relationship between the compound and its dissolved ions is one-to-one, for every mole of iron or sulfide ions that dissolves into solution one mole of FeS molecules has dissolved. This means that the molar solubility of iron sulfide is also 2.45×10^{-9} M. Remember that this is only true for compounds that have a one-to-one relationship with their dissolved ions. For a balanced equation where the coefficients are not all one, for example the dissolution of lead(II) chloride

$$PbCl_2 \Leftrightarrow Pb^{2+} + 2Cl^-$$

the solubility product relationship looks like this:

$$K_{sp} = [Pb^{2+}][Cl^-]^2$$

Since the coefficient of chloride in the balanced equation is two, then the concentration of chloride ion is squared in the K_{sp} relationship. This also affects the way that we represent the unknown value of chloride using y; there are two chloride ions for every lead ion. Therefore [Cl⁻] is represented as 2y instead of just y:

$$K_{sp} = 1.70 \times 10^{-5} = [Pb^{2+}][Cl^-]^2 = (y)(2y)^2$$

Solving for y: $1.70 \times 10^{-5} = (y)(4y^2) = 4y^3$ and $y = 1.62 \times 10^{-2}$

This calculation determines that the lead ion concentration is 1.62×10^{-2} M; but remember that the chloride concentration is twice as much (2y), so [Cl⁻] is 3.24×10^{-2} M. When these values are reentered into the K_{sp} relationship the resulting *product* is correct given significant figures and rounding (see Appendix 2).

$$K_{sp} = 1.70 \times 10^{-5} = [Pb^{2+}][Cl^-]^2 = (1.62 \times 10^{-2})(2 \times 1.62 \times 10^{-2})^2 = 1.70 \times 10^{-5}$$

However, unlike the one to one relationship between iron sulfide and its ions, the relationship *is one-to-one* for $PbCl_2$ and $[Pb^{2+}]$ but *one-to-two* for $PbCl_2$ and [Cl⁻]. Therefore, the molar solubility for lead chloride is 1.62×10^{-2} M, the same as the lead ion concentration but NOT the same as the chloride ion concentration.

QUALITATIVE SOLUBILITY RULES

So how is the solubility product relationship used in qualitative analysis? For a particular substance the solubility product reflects its solubility: *the smaller the K_{sp} is, the less soluble the compound is*. The members of the first group of molecules that we will examine in Chapter Two are traditionally separated by precipitation with hydrochloric acid (as their insoluble chlorides). The members of Group I are silver, lead and mercury(I) (Ag^+, Pb^{+2}, and Hg_2^{+2} respectively). Right now you could use the K_{sp} values for these compounds and would be able to tell that the chlorides of these three are relatively insoluble. For instance, K_{sp} for AgCl is 1.80×10^{-10}. Find the other Group I members' K_{sp} values in Appendix 1.

A short listing of common solubility rules will be useful at this point:

- Periodic Table Group IA (alkali metals) compounds are all soluble.

- All compounds containing ions of acetate, ammonium, chlorate, nitrate, and perchlorate are soluble.

- All compounds containing bromides, chlorides, and iodides are soluble except those metals in Qual Group I: Ag^+, Pb^{2+}, and Hg(I) [written as Hg_2^{2+}].

- All sulfate containing compounds are soluble except those containing Ba^{2+}, Ca^{2+}, Hg_2^{2+}, Pb^{2+}, and Sr^{2+}.

- Compounds containing Periodic Table Group IA and Ba^{2+}, Ca^{2+}, and Sr^{2+} hydroxides and metal oxides are soluble. The rest of the hydroxides and metal oxides are insoluble.

- All carbonates, phosphates, sulfites and sulfides compounds are insoluble, except for Periodic Table Group IA and ammonium compounds.

Safety and First Aid in the Laboratory

The preeminent idea that must guide your activities in the chemical laboratory is that you are responsible for your own safety. You must take care of your body (especially your eyes), your clothes, your books, and your equipment. No one else can safeguard you better than you. This, however, does not mean that you should be unaware of what is happening around you; the care that you take with your own person and work area will also help to insure the safety of others.

The second most important aspect of your safety in the laboratory is your knowledge of where important safety equipment is located. The location and safe use of fire extinguishers, eye washes, fire blankets, and safety showers will be clearly explained by your lab instructor *before* your first actual lab experiment. After this, it is your responsibility to understand and remember what you have been told about this equipment.

Your use of safety glasses is probably the most important point about safety that you can learn and practice in a laboratory. No single piece of equipment is more useful in protecting you than the safety glasses that are universally required in student laboratories. Lab instructors usually have the power to eject you from the lab if you fail to follow the laboratory policy about glasses use. Heed their warnings! The safety glasses icon that is inserted at the beginning of each experiment in this book is there to remind you to put on your safety glasses <u>before</u> you begin the experiment.

Specialized spill kits are becoming more and more common in chemical laboratories. These chemicals are designed to cleanup specific spills that could be encountered during specific kinds of experiments. A mercury spill cleanup kit or concentrated acid sorbent/neutralizer are two examples. Depending on the toxic chemicals that are used in your laboratory, appropriate spill kits will be available in the lab or from the stock room. These materials will be stocked, periodically checked, and used by your lab instructor or lab supervisor as required. If the need for a spill kit occurs, follow your lab instructor's directions exactly: These specialized materials are often most effective only if they are used in a specific way. Never use the spill kits without your instructor's knowledge, and remember to report chemical spills to the instructor.

A first aid kit is usually mandatory in chemical laboratories that are regulated by county, state, or federal agencies. These kits often include but are by no means limited to bandages, first aid tape, antiseptic wipes, instant cold packs, and a printed first aid card or instructions. Additional items for inclusion in a specialized first aid kit for a quantitative analysis lab could be poison antidotes such as EDTA or activated charcoal. It is the responsibility of your lab instructors to use these materials based upon their knowledge of the risks and benefits involved. Let them tell you what to do in the case of a medical emergency. Don't start medical procedures without specific instructions from your lab instructor or unless you have been specifically trained to do so.

The procedures in this book are offset by safety icons that help you to take the right safety precautions at the right time. The safety glasses icon has already been mentioned. Again, it is your responsibility to pay attention to the safety procedures in your area. If you have any questions about procedures or methods that the lab requires you to perform, <u>ask your lab instructor</u>. A description about the meaning of each safety icon that is used in this book is inside the front cover of this book and on the next page.

Safety Icons

This icon represents the need for **safety glasses**. It is usually used at the start of every procedure to remind students to protect their eyes.

This **poison** icon reminds the student of the danger of some of the reagents or products of their procedures. In the procedure that uses mercury compounds, for instance, this icon is juxtaposed with the textual introduction of a mercury containing reagent, $HgNO_3$.

The **corrosive** icon is used for procedures that include strong acids or bases to alert students to the danger of these reagents to their skin and laboratory surfaces in general. Any spill of these reagents should be clean up immediately and neutralized with the appropriate spill kit reagent. Any contact with the skin should be flushed away with water immediately.

The **fire extinguisher** icon alerts students that an open flame is used in the procedures, for instance, the confirmation test for tin in Chapter Three and the Group V flame tests.

An icon that alerts students to *disposal* procedures has yet to become commonly used, yet the subject of this book, environmental analysis, seems to dictate its creation and use in procedures that deal with toxic substances. In years passed, many of the reagent and products in these procedures were poured down the drain to our environment's detriment. This **safe disposal** icon will be placed where precipitate washings and experimental wastes must be disposed of appropriately. It is sincerely hoped that reminding student chemists with this symbol about safe disposal procedures will be redundant after their introduction to the environmental dangers of many of the chemicals under study.

CHAPTER ONE PROBLEMS

1. Describe the three dimensional structure of a sodium cation dissolved in water. How does this differ from a dissolved chloride ion?

2. Calculate the concentration of a solution of 300.0 grams of nitric acid (HNO_3) dissolved in water with a final volume of 2.000 L.

3. Calculate the molarity of a solution with 19.65 grams of KCl in water with a final volume of 435.56 mL.

4. How many moles of KCl are there in 50.00 mL of the solution in the last problem?

5. A solution with a concentration of 1.0 M NaCl and 10.0 mL volume is diluted to a final volume of 200 mL with deionized water. Calculate the new salt concentration.

6. The initial solution concentration in problem # 5 is 0.675 M. What is the final concentration after diluting to 200.0 mL?

* 7. A 1.5 M stock solution of NaCl is diluted serially as follows: 1.00 mL of stock is diluted to a final volume of 25.00 mL with deionized water and labelled solution A. One mL of solution A is diluted to 10.00 mL with D.I. water and labelled solution B. Finally, 2.00 mL of solution B is diluted to a final volume of 50.00 mL and labelled solution C. What is the concentration of solution C?

8. Calculate the masses of sodium chloride in all of the solution in problems 4-7.

9. Based on the common solubility rules, which of the following substances are considered soluble in water? Which are insoluble? Sodium chloride, calcium chloride, calcium hydroxide, aluminum hydroxide, aluminum bromide, sodium bromide, sodium phosphate, lead phosphate, lead sulfate, and lead chloride.

10. Calculate the solubility product of a solution of silver chloride in water that has [Ag^+] equal to 1.34×10^{-5} M and [Cl^-] equal to 1.34×10^{-5} M. Compare your answer with the K_{sp} constant for AgCl in Appendix 1.

11. Calculate the solubility product for cadmium sulfide given a cadmium ion concentration of 6×10^{-15} molar in a saturated solution.

12. Calculate K_{sp} for nickel hydroxide if [Ni^{+2}] = 3.42×10^{-6} M.

13. Calculate the molar solubility of cadmium hydroxide if the hydroxide concentration of a saturated solution above solid $Cd(OH)_2$ is 2.15×10^{-5} M.

14. Calculate the molar solubility of lead sulfate given its solubility product in Appendix 1.

15. Calculate the molar solubility of lead chloride given its K_{sp} in Appendix 1.

16. Determine the concentration of carbonate ion in a saturated solution of barium carbonate.

∗ 17. Determine the zinc(II) ion concentration in a solution in equilibrium with solid $Zn_3(PO_4)_2$.

∗ 18. Determine the phosphate ion concentration in a saturated zinc phosphate solution.

∗ 19. Calculate the molar solubility of $Zn_3(PO_4)_2$ given its K_{sp} in Appendix 1.

∗ 20. Calculate the molar solubility of silver chloride in a) pure D.I. water solution and b) a solution that already has $1.00 \times 10^{-4}\ M$ Cl^- dissolved in it before any solid AgCl is added. Assume $K_{sp} = 1.80 \times 10^{-10}$. Is there a difference? Why?

∗ 21. Why does the AgCl solubility changed with the addition of chloride anion before the silver chloride dissolves.

∗ 22. What would happen to the molar solubility of AgCl if a solution in equilibrium with undissolved silver chloride had a large amount of a very soluble salt like sodium chloride added to the solution?

Chapter 2

Toxic Mercury and Group I Cations

MINAMATA DISEASE

Beginning in late 1953 on the island of Kyushu in Japan, it became clear that something was very wrong. People who lived around Minamata Bay on the western shore of the island were getting sick. Adults, children, and newborn babies who lived around the bay were showing signs of severe health problems such as numbness, distortion of vision, loss of muscle control, speech and hearing impairment, unconsciousness and even death. Within the next seven or eight years over one hundred people were identified as suffering from the effects of the, at first, mysterious aliment dubbed Minamata Disease. The one thing that all the victims had in common was that they lived on the seafood caught in and around the bay. Indeed many of the victims were fisherman and their families. By 1992, over 1200 people had died of damage related to this disorder.

Minamata disease is caused by mercury poisoning; however, though mercury is a naturally occurring element (like all the elements that we will discuss in the book), it is not normally found in dangerous doses in the unpolluted environment and is very seldom found in toxic levels in fish. Where did the mercury that was poisoning the people around Minamata Bay come from? A chemical plant near the bay used mercury compounds (mercuric chloride and mercury oxide) as a catalyst in the production of other chemicals. The catalysts were usually carefully handled by the plant operators because of the known toxicity of mercury, but four times a year a relatively large amount of mercury was discharged into the bay during plant cleaning. This mercury had been collecting in the mud in the bay and was **bioconcentrated** by the fish and shellfish who lived in the bay. This means that the fish built up in their tissues some compounds containing the metal by coming in contact with mercury in the mud on the bay's bottom or in the waters of the bay. After very careful studies by a number of Japanese scientists, it became clear that some of the waste forms of mercury that were being discharged into the bay were even more toxic than the mercury used in the catalyst itself.

One of the industrial processes in which mercury was used converted inorganic mercury (that is, present in molecules containing no carbon) to organomercury by bonding it to a methyl group (CH_3—). The problem with organic mercury compounds is their solubility in the fats in our bodies. Since they are fat soluble they can accumulate in our tissue and brain and thereby present more danger to humans than the more soluble mercury salts which can be more quickly flushed out. Hence, organomercury compounds such as methyl mercury chloride (CH_3HgCl) are many times more toxic than inorganic mercury compounds such as mercuric chloride ($HgCl_2$).

It is interesting that the Environmental Protection Agency still uses an analytical lab procedure that relies on mercury as a catalyst. Although a replacement has not been found so far, less toxic metals are under scrutiny as successful replacements.

Mercury Pollution in the Amazon

The Minamata Disease episode has, unfortunately, been repeated at other sites in the industrialized world; however, even developing countries like Brazil are also at risk. Native gold miners in the Amazon rain forest are using mercury to separate gold from river silt. The mercury containing wastes from this process are being dumped into the Amazon River system or released as vapors when a mixture of gold and mercury is heated with a blow torch to reclaim the gold. The amount of mercury that has been released into the environment may be over 1500 tons in the past few decades and evidence of people with mercury poisoning has already been reported by a Brazilian doctor.

60 Minutes, Amalgams, and Mercury

Though we in the United States may feel insulated from the types of pollution episodes described above, there is at least one form of mercury with which, over our life times, most of us will probably come in contact. Fifty percent of a typical dental **amalgam** filling is mercury. When this metal is mixed with silver and other metals like tin, copper, and zinc in trace amounts it forms a hard material that is packed into the hole in a tooth that has been drilled out by a dentist. The result is a very hard surface that protects the tooth from further decay. It also creates a hard surface, particularly in molars, on which to chew. In the United States about one hundred million amalgam fillings a year are installed.

Though there is a good probability that some mercury is vaporized from amalgam fillings when we chew, it is also probable that for most people the amount of mercury absorbed by the body during this process presents no health hazard whatsoever. This aspect of "oral pollution" has been hotly debated in recent years. Strongly inflamed by a program on television at the beginning of the decade ("60 Minutes", December 16, 1990), different groups have argued that the toxic effects of mercury leaking out of dental fillings, either directly into the body or vaporized by chewing and then inhaled, have lead to the breakdown of immune response and caused arthritis, leukemia, and multiple sclerosis. These claims, for people without specific mercury allergic reactions, have not yet been scientifically validated. And while it is true that people who come in contact with mercury compounds used in the amalgam process, like dentists, have a higher measurable urine concentration of mercury, the statistical connection between, for instance, the number of amalgam fillings and blood or urine content of mercury has not been made. There is a good chance that we consume more mercury from our normal everyday contact with environmentally safe water, food, and air where it is present in very low amounts, than from dental fillings; however, studies in this area are continuing and some people have had their amalgam fillings removed in efforts to improve their health.

SAMPLING FOR MERCURY IN THE ENVIRONMENT

Aqueous solutions and biological samples containing mercury are often oxidized to the mercury(II) oxidation state before analysis to insure that all the mercury present is dissolved and available for analysis. Strong oxidizing agents like hydrogen peroxide or permanganate are usually used to accomplish this. Note that the Group I Qual (short for qualitative analysis) scheme below is specifically for samples in which Hg(I) is known to be present.

It is interesting that oxidized mercury is often chemically reduced back to elemental mercury for a number of instrumental analyses. Metallic mercury has a low enough boiling point to allow it to be evaporated and thereby collected as the metal for analysis. Elemental mercury's low boiling point (bp = 357 º C) is a very unusual property and not approached by any other metal. Vice President Al Gore writes that the release of toxic mercury from the increasing use of incinerators may, in fact, be the greatest mercury threat in the U.S. today.

QUALITATIVE ANALYSIS OF MERCURY AND GROUP I CATIONS

The first cation group in the classical qualitative scheme (from here on called **Group I**) contains silver(I), Ag^+, mercury(I), Hg_2^{2+}, and lead(II), Pb^{2+}. It is important to note that the formula for the mercury(I) cation is Hg_2^{2+} not Hg^+ as you normally would expect. The Hg^+ ion is unstable in solution and forms the **dimer** Hg_2^{2+} instead.

We will use the solubility of these species (as various chemicals complexes) as a tool for their separation from a mixture; but first we will make a precipitate of mercury chloride from one of its only soluble forms, mercury(I) nitrate, so that you can see a clear example of Hg_2Cl_2.

Precipitating Hg_2Cl_2 from a solution of $HgNO_3$

Make sure that you recognize all of the safety icons that are used in this book. They are inserted in procedures where they are needed to remind you to be careful!

Put 2 mL of 0.1 M solution of $HgNO_3$ [also written $Hg_2(NO_3)_2$] in a test tube. To this solution slowly add 0.5 mL of 6 M HCl. Using an eye dropper this will be about 10 drops. As you add the acid you will see the instantaneous formation of mercury(I) chloride. This white precipitate will form as the concentration of Hg_2Cl_2 exceeds the solution concentration determined by the solubility product. After adding 10 drops, hold the top of the test tube

with one hand and with the other hand stir the solution by gently tapping the test tube with your finger. Wait for the solid that has formed to settle and then add one more drop to make sure that the precipitation is complete. If more precipitate forms add more acid until no more precipitate forms. Label the test tube with a wax pencil or paper label and put your sample of Hg_2Cl_2 aside in your test tube rack.

The significant differences in solubility between Hg_2Cl_2, $AgCl$, and $PbCl_2$ in hot water solution is the key to separating these species from a solution containing all three. Lead(II) chloride is more soluble in a hot water solution than Hg_2Cl_2 or $AgCl$. If a mixture of these three precipitates is heated to water's boiling point, the lead compound will redissolve and can then be decanted (poured off) while the solid Hg_2Cl_2 and $AgCl$ remain behind in the test tube. The separation can be helped by centrifuging the mixture before decantation.

Separation of Hg_2^{2+} and Ag^+ from Pb^{2+} using solubility differences

Put a medium size beaker two thirds full of tap water on a hot plate and bring it to a boil. This is your hot water bath.

Add 10 drops of 6 M HCl to a test tube containing 2 mL of a solution containing nitrates of Hg_2^{2+}, Ag^+, and Pb^{2+} (again 0.1 M in each). Allow the solids that form in the solution to settle. After the first ten drops you will notice that the next drop of acid probably continues the precipitation. Slowly continue adding drops of HCl until precipitation stops. Stir, as before, with a finger tap on the test tube between every few drops.

After the precipitation has finished, add two more drops of acid and then add 2 mL of distilled water to the test tube and place it in the hot water bath for 2 minutes. Stir the test tube every thirty seconds with a few finger taps. After 2 minutes of heating, carefully **decant** the liquid above the solid (the **supernatant**) into another test tube or first centrifuge the test tube with the solid and then decant the supernatant. Do not wait more than a few seconds to perform this step after you have removed the heated test tube from the water bath; the success of this test depends on the difference in solubility of Hg_2Cl_2, $AgCl$, and $PbCl_2$ *in hot water*.

The original test tube now contains a mixed precipitate of Hg_2Cl_2 and $AgCl$ similar to the sample of Hg_2Cl_2 in the test tube rack. Compare these solids.

The supernatant that you decanted may slowly precipitate $PbCl_2$ again as it cools. You may be able to speed this up by cooling this test tube under running water or by putting it into an ice bath for a few seconds. If the precipitation still does not occur and you want to definitely confirm the presence of lead, add 0.3 M potassium chromate (K_2CrO_4) dropwise to the supernatant until a yellow precipitate forms. This precipitate is lead chromate ($PbCrO_4$) and confirms the presence of lead in your supernatant from the previous step. Label and set this test tube containing $PbCrO_4$ in your test tube rack.

Separation of Ag^+ from Hg_2^{2+}

The final member of Group I, Ag^+, is also precipitated as its chloride in cold water. This means that if it is also present in the original solution (it was), it will be mixed with the Hg_2Cl_2 precipitate left over after the decantation of the supernatant that contained lead (above). The presence of silver in this precipitate can be confirmed by 1) selectively re-dissolving Ag⁺ using an ammonia solution (that is, dissolving silver as the Ag(NH$_3$)$_2$⁺ complex), 2) separating it from the Hg$_2$Cl$_2$ solid as a new supernatant, 3) re-acidifying that solution to neutralize the complex, and then 4) checking for AgCl.

Take the precipitate left over after the lead decantation step and wash it to remove any traces of lead. Add a volume of water equal to the volume of the precipitate in the test tube. Heat the test tube in a water bath for 1 minute. Stir a few times with a few finger taps, and then centrifuge and decant the supernatant. Repeat this wash with an equal volume of water followed by heating and centrifuging. These washings will remove any last traces of lead that may be present. *Remember to put your wash solutions into the proper disposal container.*

Take the washed precipitate and add 2 mL of 6 M NH$_3$. This solution will turn black or gray if mercury is present (and it should be if all has gone well). This is due to the presence of two mercury species (elemental mercury and HgNH$_2$Cl). Centrifuge and pour off the supernatant into another test tube. Label and store the mixed mercury precipitate in your rack.

Since you started with a mixture of all three Group I cations and you have now removed and identified lead(II) and mercury(I), the only metal cation left in this supernatant is silver. How can we confirm this?

The presence of silver can be confirmed by neutralizing the basic ammonia solution with nitric acid and then checking for insoluble silver chloride. Add 12 M nitric acid dropwise until precipitation occurs. This precipitate is silver chloride, and a white precipitate that is present after these series of steps is considered a positive test for silver.

Make sure that the solutions and precipitates that you have produced in your analyses are disposed of in the appropriate container. *If you are not sure where this container is, ask someone who knows!*

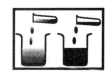

Flow chart for analysis of Group I cations

The experimental procedures for the separation and identification of members of a particular cation group (or any systematic analysis scheme) can be represented using a flow chart. This is a visual tool to help you logically organize the process of separation. It is just like a map except the "places" are the chemical formulas for dissolved or precipitated ions and the streets are logical progressions from one form to the next.

There are many ways to put together a flow chart for a particular Qual group and you will be asked to create a few in the chapters to come. One way to represent the qualitative analysis scheme of the Group I cations is on the next page.

Group I unknowns

Get a test tube containing a Group I unknown from your instructor. Record the unknown number if this is applicable. Divide the solution into at least two portions so that you can repeat your separation scheme from the beginning if it becomes necessary. Assume that your unknown contains all the members of this group that we have studied. Write down the steps of the scheme as you perform them making careful notes about the success of each confirmation step. Repeat the known confirmation tests if you are allowed to and if the need arises because of an indistinct result. Report to your instructor your separation scheme and the Group I members present in your unknown.

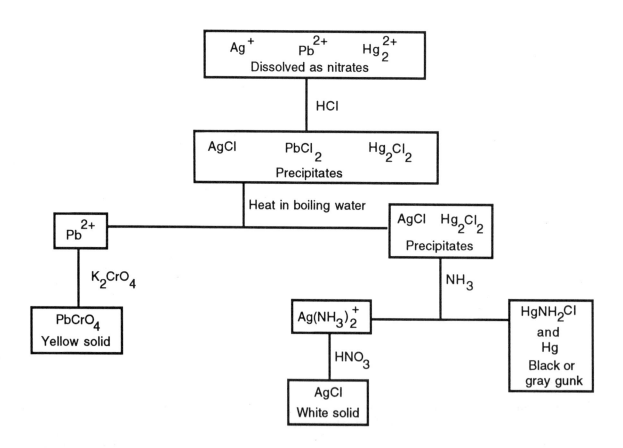

Flow Chart for the Qualitative Analysis of Group I Cations

INSTRUMENTAL ANALYSIS OF MERCURY

The methods that you have just used in the previous section to detect mercury and the other members of Group I are called **wet chemical methods**. They involve the classical tools of the analytical chemist—beakers, test tubes, and the centrifuge—and the classical procedures of the analytical chemist—filtration, decantation, and centrifuging (centrifugation). These tools are used to handle solutions, complexes, and precipitates of chemicals that you want to identify in a sample. And as you have just learned, if you know the color, solubility, and reactions of cations in a group and specific complexes of those species, you can successfully detect individual members of a complex mixture. Chemists have been doing these types of analyses for over one hundred years. Applying the term classical to these methods is an historical classification and is used because qualitative methods chronologically precede the instrumental methods that we are about to discover.

Important requirements for the classical Qual scheme are

1) the concentration of the group members must be high enough to give definite positive tests

2) interferences must be kept low enough to not cause false positives—a positive test for a species that is not present—or false negatives—a negative test for a species that is present.

In actuality, if the concentration of a chemical species, for instance mercury, is very low or if it is dissolved in solution along with a strongly interfering species, problems arise. The separation scheme becomes much more complicated and can involve many additional steps and different procedures than just those that you have practiced here—procedures that can be very time consuming and subsequently expensive.

For toxic metals like mercury (and others that you will be introduced to later) concentrations well below those easily detected using wet methods can be dangerous. For instance, the Food and Drug Administration (FDA) limits the highest amount of mercury that can be detected in fish for sale in the United States as 1 part per million. This is 1×10^{-6} grams of Hg detected in 1 gram of fish—a very small amount of mercury indeed. For routine methods of mercury determination, something else besides time consuming, relatively insensitive wet chemical methods is necessary. Enter instrumental chemical analysis.

Instrumental methods of analysis measure physical properties of atoms or molecules in order to detect them. But unlike the colors of solution or solubilities used in our Qual scheme, the characteristics detected by instrumental methods are things like atomic or molecular light absorption or emission, conductivity, or chemical potential to list just a few.

The ability of a particular atom, for instance mercury, to absorb light is a very powerful method of detection. This physical characteristic of all atoms of mercury allows us to very sensitively determine the presence of these atoms in mixtures containing other atoms or molecules that would normally interfere with our wet techniques. In addition, because the technique is so sensitive, this test to say yea or nay about the presence of toxic mercury can be applied to very little small samples and very small amounts of mercury. Testing to determine if a sample of swordfish meets the FDA requirement of 1×10^{-6} grams of Hg detected in 1 gram of fish (1 part mercury in a million parts of swordfish) can be accomplished in a very few minutes and with great accuracy using the instrumental method called atomic absorption spectrometry.

Atomic absorption spectrometry

Just as the number of protons in the nucleus of an atom (the atomic number) determines elemental identity, the number of electrons and the spacing between those electrons in the atom's **electronic shells** and subshells is also a characteristic that is particular to each kind of atom. Every neutral mercury atom, for instance, has the same number of electrons in the same shells no matter where the mercury comes from—dental amalgam, industrially contaminated shellfish, or soil from the bottom of the Amazon river. If we can find a way to detect something about mercury atoms that depends on the existence and spacing of those electrons, we have a means to detect mercury itself.

The relationship between the electronic shells of atoms is actually a measure of the relative energy of those shells. Electrons can move between the individual shells only by absorbing or emitting energy. Furthermore, the amount of energy that will cause movement of a particular electron in a particular electronic shell to another shell (called an electronic transition) is very specific. This means that the energies that electrons in a particular atom will absorb or emit are very well define or **quantized**.

The energy of light waves changes with wavelength (think of differences in color as differences in wavelength—blue light is a different wavelength and higher in energy that red light, for instance). If we want to selectively excite the electrons of a mercury atom—that is, purposely cause an electronic transition—we can chose the specific energy of the mercury atom's electronic transition by using a particular wavelength of light. In the case of mercury this is usually 253×10^{-9} meters (253 nanometers). In effect, we can "tune in" to the wavelength of mercury's available electronic transitions.

Atomic absorption spectrometry (AAS) uses this process to detect mercury. Light of a specific energy (think wavelength or color) is shined through a gaseous sample containing mercury atoms. The mercury atoms that are present in the sample absorb some of the light causing the particular electronic transitions available to mercury's electrons. The amount of light *exiting* the sample is therefore missing some of the light of this wavelength because it was absorbed by the sample's mercury atoms. If we measure the difference between what goes into the sample and what comes out and we see some missing light, voilà! Mercury has been detected. If no mercury is present in the sample, the amount of light entering and exiting the sample is equal and the qualitative question "Is mercury present?" is answered "no."

Another useful feature of this technique is that the *amount* of light that is missing on the exiting side of the sample depends of the amount of mercury in the sample. Not only has qualitative analysis occurred—answering the question "was mercury present?"—but **quantitative analysis**—"how much mercury was present?"—has also been accomplished simultaneously.

A simple atomic absorption spectrometer

So how are these instrumental requirements met? The illustration that follows is a schematic of all of the necessary parts for a simple atomic absorption spectrometer. It shows a light source, a sample vaporizer, a light separator, and an "electric eye."

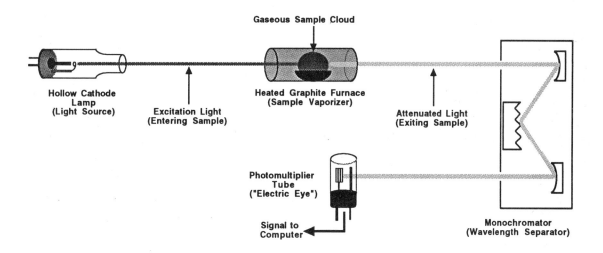

Figure 2.1 Schematic diagram of an atomic absorption spectrometer.

The particular wavelength of light required for the analysis of mercury, for instance, is obtained using an electric lamp that produces almost exclusively the wavelength that only mercury atoms can absorb. This device is usually a **hollow cathode lamp** made specially for mercury detection. The AAS analysis for each element requires a hollow cathode lamp made for that element. There are, however, multielement lamps available with two or three element sources in one lamp but the AAS instrument must be adjust individually ("tuned") for each element before analysis.

In AAS, a sample is vaporized by feeding it into a flame or by placing the sample in a graphite tube that is heated to temperatures high enough to cause vaporization and atomization (> 2000 º C). The example of the graphite furnace is shown in our illustration (instead of the flame) because many environmentally important biological samples containing mercury are analyzed in this way. However, flame AAS is still the most commonly used atomizer/vaporizer.

If mercury is present in the sample's vapor, the light passing through the sample is decreased (attenuated). The light exiting the graphite furnace passes through a device that separates any extraneous or interfering wavelengths that may be produced by the hollow cathode lamp or by the hot graphite furnace. This light separation device is called a **monochromator** and it selects only the wavelength characteristic of mercury for passage on to the electric eye—all other wavelengths are blocked. Finally, the light exiting the monochromator is directed onto a light measuring instrument called a **phototube**. The job of this device is to turn light energy into electrical energy that can be sent to a computer. Actually more sensitive **photomultipliers** are often used in this instrument instead of the less sensitive phototubes but the principle of the conversion of light to electrical current is the same.

The photomultiplier does the same thing that your eyes do. Both of these sensitive "instruments" act as **signal transducers**; they convert one form of energy into another. The light reflecting from this page enters your eye and is focussed onto your retina by the lens in your eye. The retina converts the light's photons into electronic nerve signals that are sent to the brain by the optic nerve. Just as the phototube or photomultiplier tells the AAS about light exiting the graphite furnace, your eye tells you something about the characters on this page and whether or not this sentence ends with a period

Most instrumental methods of analysis use chemical standards as a means of calibrating the instrument. **Calibration** is a systematic way of determining how much electronic signal corresponds to a specific amount of the element of interest. In the case of mercury, this might correspond to the relationship of photomultiplier output (usually in units of electrical current, amps) versus different amounts of mercury in a sample or standard mixture—more mercury will generate a larger signal. This mathematical relationship can be represented in the form of an x-y plot.

Since calibration curves are common to most instrumental methods we will come back to this analytical tool in the next chapter.

AAS EXPERIMENT— DETERMINATION OF MERCURY

The AAS determination of mercury is routinely based on the detection of the Hg absorption signal at 253.4 nm using a lean acetylene/air flame optimized for Hg detection. This experiment describes a means of optimizing the absorption signal of an AAS instrument while aspirating a solution of mercury with a known concentration. The results are plotted as absorption signal versus various parameters: flame mixture, slit width, and three different burner heights. Alternatively an iron standard may be substituted for environmental considerations since the instrument's operation is essentially identical for each element.

1. Prepare the instrument following the manufacture's instructions; tune the monochromator to the 253.4 nm mercury line and set the slit at the suggested width. [The iron line is 248.3 nm.]

2. Get a mercury (or iron) solution of known concentration from the instructor or make a 50 ppm mercury standard by dissolving 0.006 grams of mercury(II) oxide (or 0.015 grams of iron(III) chloride) in 5 or 10 mL of 50% hydrochloric acid solution in a 100 mL volumetric flask. After the solid is completely dissolved, add water until the solution level meets the line etched in the neck of the flask. Insert the stopper and invert the flask repeatedly to mix the solution completely.

3. Have the instructor light the flame using the fuel/air mixture specified by the manufacturer. Make sure that the exhaust hood or flue is turned on. Do not operate the flame without adequate ventilation. The lighting of the flame is usually microprocessor controlled and therefore automated. The computer won't light the flame if safety conditions are not met: For instance, if the fuel/air mixture is incorrectly adjusted or the burner chamber's safety window is not in place the computer reports a lighting error.

4. Beginning with the parameter settings detailed by the manufacture, aspirate the standard metal solution and record the absorption reading after 5 seconds. Stop aspirating the sample and carefully adjust the flame mixture by *leaving the air flow untouched* and increasing the acetylene flow by 20 % using the instrument's acetylene gauge. Aspirate the sample and record the new absorption reading at this rich mixture after 5 seconds. Stop aspirating the sample and repeat the fuel adjustment only this time decrease the mixture to 20 % below the manufacture's settings. Record the absorption reading at this lean setting after 5 seconds of solution aspiration.

Plot absorption readings (y-axis) versus the three fuel flows (x-axis in cc/min or psi). Label your plot **fuel mixture optimization**.

5. Leaving the fuel mixture at its optimum setting and the slit width at the manufacturer's suggested width, aspirate the standard metal solution and after 5 seconds record the absorption reading. Adjust the slit width to one setting smaller and after 5 seconds record the absorption reading. Repeat at one setting larger than the manufacturer's suggested setting and after 5 seconds of sample aspiration record the absorption.

Plot absorption readings versus slit width. Label your plot **slit width optimization**.

6. Repeat this experiment at three different burner heights. Adjust the vertical burner position one full turn of the adjustment knob each time. Make sure that the burner head does not actually block the lamp beam itself.

Plot absorption versus low, medium, and high burner positions. Label your plot **burner height optimization**.

7. Make sure that the solutions that you have produced in your analyses are disposed of in the appropriate container. *If you are not sure where this container is, ask someone who knows!*

CHAPTER TWO
PROBLEMS

1. What are the physical effects of Minamata Disease and what chemical element causes this disease?

2. When and where were the effects of Minamata Disease first recognized?

3. What is the definition of bioconcentration?

4. Why are organic mercury compounds more dangerous to us than purely inorganic mercury?

5. What is an amalgam and why is it discussed in this chapter on mercury?

6. Are dentists more likely to have higher or lower mercury content in their blood than the normal population? Why?

7. Why are the solubility products of the Group I nitrates not included in the appendix that lists solubility products?

8. What is the purpose of using hydrochloric acid when precipitating the Group I cations?

9. Using the solubility product table in the appendix, list the Group I chlorides in order of increasing solubility.

* 10. Draw a flow chart clearly detailing the reagents that you will use and the results of each step that you propose for the separation of Group I cations if you started with an unknown solution containing only the nitrates of silver and lead.

11. How can you differentiate between mercury and lead ions in solution?

* 12. Mercury(I) undergoes a disproportionation reaction in the presence of ammonia. Using the products noted in the procedure write the reaction of mercury(I) chloride with ammonia. Note the oxidation state of both mercury products.

13. How can you identify silver cations in the presence of mercury(I) cations?

* 14. Using a very sensitive analytical balance, describe how you could determine the *amount* of silver in a sample of river water that is delivered to your laboratory.

15. Write balanced complete ionic equations for the reaction of dissolved silver nitrate and dilute hydrochloric acid.

16. Write down the net ionic equation for the reaction in problem 12.

17. Write down the balanced ionic equation for the reaction of an aqueous solution of potassium chromate and dissolved lead nitrate.

18. Write down the net ionic equation for the reaction in problem 14.

* 19. What are the limitations of classical qualitative methods?

20. How do the limitations that you noted in question 19 affect analyses in modern analytical laboratories in regards to trace analyses of toxic metals?

21. Compare the chemical complexity (number of compounds involved) in the samples that you analyzed in this chapter's procedures and those that you would obtain from the Amazon River basin. Are environmental samples more or less complex than your laboratory samples? Why?

22. How do you think that the growth of electronics and computer industries have affected the availability of analytical instruments in the laboratory?

23. How do you think that the availability of instrumentation in the analytical laboratory has affected the speed and expense of environmental analysis?

24. All nuclei of a particular element have the number of what particle in common?

25. How many electrons do all neutral mercury atoms have in their electron shells?

26. What happens when an electron in an atom's electronic shell absorbs energy?

27. What happens when an electron in an atom's electronic shell emits energy?

28. For transition between two different electronic shells, what do the photons absorbed and the photons emitted have in common?

29. What is the relationship between the color and the wavelength of light?

* 30. Do you think that the energy of *emitted* photons could be used as an analytical tool the way that energy of *absorbed* photons are used in atomic absorption spectrometry?

31. What is the name of the "electric eye" used in AAS?

32. Why is the high temperature required in the graphite furnace used in AAS?

33. What is the definition of the word monochromator and what is its purpose in the AAS instrument?

* 34. How can the *amount* of mercury present in a sample be determined using AAS?

*35. Do you think that interfering compounds are as much of a problem in AAS as they are in the classical Qual scheme? Why?

36. Write down the definition of calibration in regards to an AAS instrument.

*37. How is the AAS instrument calibrated?

38. How could you make up calibration standards of soluble mercury compounds?

*39. Why isn't there a single lamp available for all AAS elemental analyses?

*40. What are the possible spectral interferences from a multielement lamp?

*41. What do you think happens to the AAS elemental lamps as they age?

Chapter 3

Environmental Cadmium

SOURCES IN OUR ENVIRONMENT

Unlike mercury, which has a long history of use by humans stretching back into antiquity, cadmium was just an unused contaminant of zinc smelting until it finally found uses during the second part of this century. **Anodizing** is an electrical process where by a metal can be electrically plated onto another metal to protect it. The plating usually takes place in a bath of soluble metal salts. The metal part is dipped into the bath while under electrical potential (connected to a power supply). Therefore, the process is also called **electroplating**. Metal parts anodized with cadmium (as with zinc) are very resistance to attack and will last longer and out perform equivalent parts that are uncoated.

All parts anodized with zinc also contain some cadmium. The cheaper the zinc grade used for plating the more cadmium is present as an impurity. Many incidences of "zinc poisonings" were probably cadmium poisonings in disguise for two reasons: "Pure" zinc is almost always contaminated with cadmium, and zinc is substantially less toxic than cadmium. In fact, zinc is actually a trace metal that is required by the body for health.

The pollution controls on electroplating have not kept up with its widespread use in the United States. Pollution of rivers and estuaries around urban centers (where electroplating shops are concentrated and waste metals are illegally dumped into sewers and storm drains) has been increasing for at least the past three decades.

The uses of cadmium in our society are not limited to electroplating metal parts, though on a mass basis this is probably where most cadmium has been used (and spilled). Because of the relative energies of the electrons around the cadmium nucleus (the electron shells), this toxic metal is used in semiconductors too. A **semiconductor** is a solid material that can be used in an electronic device such as electronic circuits, microprocessors (computers), solar cells, or even semiconductor **lasers**.

For instance, a semiconductor can be made by the combination of elements like lead and sulfur with cadmium. The result is a crystalline material that has specific physical characteristics for the job at hand—in this case, a diode laser. When electric current is flowed through this particular semiconductor the result is the release of a high intensity light beam with a very pure color; this is a laser. This particular laser is quite small (0.4 by 0.2 mm) and therefore requires only a very small amount of cadmium to manufacture. Note that this is unlike a large machine part which may have its entire surface electroplated in an anodizing bath.

Another use of cadmium (and a potentially major source of pollution) is nickel-cadmium (NiCd) batteries. Because these kinds of batteries can be sealed and are so small, they are an excellent source of electricity for electrical devices that have only a limited space (or weight) available for a battery. Calculators and portable ("laptop") computers are good examples of devices with these requirements. In 1993, the fastest growing segment of the computer market is portable, battery powered computers. Though less environmentally hazardous batteries are surely coming, millions will be sold before they are replaced. The question is "What happens to the NiCd batteries once they are exhausted?" What have you done with NiCd batteries that you removed from a watch or calculator or toy?

Just as the most common source of mercury that we are routinely exposed to is probably dental amalgams (besides the minute ever-present background), the most common source of cadmium may be another wide spread component of our society—cigarette smoking. Each cigarette, besides releasing other toxic molecules like carbon monoxide, formaldehyde, and radioactive polonium, releases approximately 1 microgram of cadmium per cigarette. This poison is probably in the form of solid **particulates**. These small particles are either trapped in the smoker's lungs or are deposited as second-hand smoke in the vicinity of the smoker and thereby endanger others present.

PHYSICAL EFFECTS OF CADMIUM

In Japan, once again, an industrial pollution accident occurred that involved symptoms that, like Minamata Disease, were given a specific name. Post-menopausal women poisoned by lead and cadmium from a metal smelter were referred to as the victims of *itai-itai* disease, literally translated as ouch-ouch disease. These victims were poisoned by the water from a river used for irrigation and drinking water that had been contaminated by the ore waste from the smelter (this form of solid waste is traditionally called **ore tailings**). They suffered from deformities, kidney damage, bone breakage due to calcium loss, and sometimes death. The soil and river bottom near the smelter will be contaminated for a long time to come since the contamination was so extensive.

Besides the effects of large doses of cadmium exhibited by victims of itai-itai disease, what physical effects does cadmium have? Since cadmium is retained by the body after absorbing it, cadmium can build up over long periods of exposure. Like some of the mercury in the body, cadmium is stored in the liver and kidneys; however, zinc and cadmium apparently compete for sites in the body, and more cadmium in the body can therefore mean less zinc. Since zinc is a necessary dietary component and cadmium a poison, this competition can be a life and death affair.

High blood pressure, kidney damage, and possibly emphysema are results of long-term and low-concentration doses of this metal. Tests on people who died with high blood pressure usually show an elevated cadmium level (or a reduced zinc level) in their kidneys compared to those who died of other causes. In tests designed to duplicate this relationship in the laboratory, dogs, rabbits, and rats all had measurably higher blood pressure when fed diets containing cadmium.

SAMPLING FOR CADMIUM IN THE ENVIRONMENT

The methods of sampling for cadmium in the environment often involve the separation of organs from animals and the analysis of these organs individually for cadmium content. For instance, since ingested cadmium concentrates in the liver and kidneys of animals, these organs are often removed, dissected, and, before analysis, weighed slices are either extracted with strong acids or treated with chemicals that capture cadmium as a soluble complex. This last process is called **chelation** (probably derived from the Greek *chela*—claw). It involves the use of a solution containing reactive molecules that can surround the metal species and dissolve it away from the animal tissue matrix. The solution of cadmium chelate is then analyzed.

In the case of airborne cadmium sampling, for instance dust particles in the air near a smelter or a tailings heap, the air that contains dust is pulled through a filter with small pores to catch the particulate matter. After sampling a known volume of air, the filter is washed with an acid or chelating agent and the resulting solution is analyzed for cadmium.

CLEANING UP CADMIUM POLLUTION

The viability of pollution cleanup in our society is unfortunately almost exclusively an economic proposition. This means that the society does not put unlimited value on the cleanup of dangerous chemicals. In effect, the danger to people from pollution is balanced by the danger to society's pocketbook since there is not an unlimited amount of money available to accomplish this job, or indeed almost any job that most of the society considers important. Factor this along with the interest that some polluters may have in minimizing their responsibility (which can be construed as maximizing their profits) and the cost to our society of other concerns—defense, international aid, and entitlements, and you have approximated the complexity of this situation in our country. The intricacy of the pollution situation in other countries, for example newly independent Poland or Hungary is completely beyond the scope of this book, maybe any book that has been written so far.

One of the major problems with the cleanup of toxic materials spilled into the environment is that sometimes they are doing less damage to the environment where they are than if they were moved to another site. This is, of course, little consolation to people who are close to the pollution site and are acutely affected; however, toxic material like cadmium, precipitated in the mud of New York harbor will probably be better left alone than dredged up. Here's why: when precipitates in natural (and complex) systems are stirred up they often redissolve to a certain extent. A Polish professor visiting the United States in 1991 once said to this author that a very polluted river flowing through Gdansk, Poland was dredged periodically. Since people got drinking water from the river, when the dredging occurred, people got sick—the physical agitation of the dredging process caused the release of toxic material from the mud on the river's bottom.

Another aspect of this problem involves where to put the material that you would remove from a site like, for instance, New York Harbor or the Houston Ship Channel. Although the people near this site want the toxics removed, the people living somewhere else are not interested in having someone put thousands of tons of mud containing toxic metals nearby. This is an honest human response that has been dubbed the **NIMBY** syndrome (Not In My Back Yard), *which may be considered the human equivalent of maximizing your profits.*

These arguments may seem like a way to excuse our society from its requirement, moral and otherwise, of keeping the environment clean or cleaning up our mess; however, this is just a pragmatic view of a very complex situation—a situation that must be examined closely and handled carefully. Clearly, not polluting our environment in the first place is the *best* solution to pollution.

QUALITATIVE ANALYSIS OF CADMIUM AND GROUP II CATIONS

There are eight members of the Group II cations in the classical scheme. This number could be expanded to eleven if you count multiple oxidation states for particular elements. We will confine our Group II work to bismuth(III), Bi^{3+}, cadmium(II), Cd^{2+}, copper(II), Cu^{2+}, and tin(IV), Sn^{4+}. The final confirming test for cadmium usually involves the creation of a yellow CdS precipitate. This is made by passing H_2S, hydrogen sulfide, through a solution containing dissolved Cd^{2+}. We will perform this step first with a solution containing only Cd^{2+} using a method that avoids having to make a gaseous H_2S bubbler. Instead we will use a reagent called thioacetamide that releases H_2S in water solution in the presence of dilute acid (or base) as a source of H_2S. During these procedures using thioacetamide, remember that H_2S is slowly being evolved and that this gas is toxic. Perform your analysis under a hood. You will be able to recognize the smell of hydrogen sulfide instantly and should avoid breathing it.

Precipitating CdS from a solution of $CdCl_2$

Put a medium-size beaker two-thirds full of tap water on a hot plate and bring it to a boil. This is your hot water bath.

To 2 mL of a solution of 0.1 M $CdCl_2$ in a test tube add 2 drops of 3 M HCl. Do not add extra acid. To this acidic solution add 10 or 15 drops of 1 M thioacetamide solution using an eye dropper. Heat this test tube in your hot water bath until you see the yellow CdS precipitate. Remember that H_2S is slowly being evolved and that this gas is toxic. Perform your analysis under a hood. After the precipitation is complete, heat for a minute or two longer and then place your sample of CdS in your test tube rack. If a stopper is available, gently insert a small stopper in the mouth of the test tube before you put it in your rack.

Precipitating the Group II cations as insoluble sulfides

To approximately 3 mL of a solution containing soluble salts of all of the four selected members of this group add 8 mL of ammonium sulfide solution (10 % $(NH_4)_2S$ by wt.). Heat this test tube in a hot water bath for five minutes. Watch out for excessive foaming—momentarily remove the test tube or cool down the hot water bath if it is becomes necessary.

After heating, add 1 mL of 1 M NaOH solution to this test tube. Three of the members of this Qual group will precipitate as insoluble sulfides. The fourth, tin(IV) sulfide will stay in solution in this basic solution. Centrifuge and collect the precipitates from this step. Put the (tin-containing) supernatant in a test tube to be analyzed for tin. Clearly label this test tube.

Wash the collected precipitates with 2 mL aliquots (portions) of cold distilled water and 1 mL of 1 M NaOH by adding the water then the base, stirring the solution and then centrifuging. Repeat this wash once. Dispose of the wash water in the appropriate container. Label the final test tube containing your Group II insoluble sulfides.

Detecting tin(IV) with mercury

Tin is traditionally detected by its reaction with aluminum to produce tin metal (Sn^o) and then further reaction with HCl to form Sn(II). The presence of tin(II), and therefore the original Sn(IV), is confirmed by its reaction with colorless mercury(II) chloride to produce metallic mercury and mercury(I) chloride. We will perform this test and a flame test using a known solution of tin chloride.

Acidify the tin(IV) sulfide containing supernatant from above by adding 10 drops of 6 M HCl. Under a hood put this solution in the boiling hot water bath for

1 minute and then remove and add 10 more drops of acid. Repeat these steps until 40 drops of acid have been added. Some of this solution can be saved for the flame test below by pouring off about 1 mL into a test tube labeled accordingly.

Add a 1 inch piece of 24 or 26 gauge aluminum wire to the test tube that was repeatedly acidified with HCl. Heat the test tube in the hot water bath until the wire dissolves completely. Heat further until almost everything in the solution dissolves (some stubborn solid may remain). You may need to add a few drop of 6M HCl to help complete the dissolution.

To prevent an interfering reaction with the air, *quickly* cool the test tube with tap water and add 2 drops of 0.1 M $HgCl_2$ [mercury(II)] solution to the test tube. The slow formation of a gray or white cloudy precipitate confirms the presence of tin.

Detecting tin(IV) by a flame test

Get a small volume of standard tin salt solution in a test tube. Fire polish a Nichrome wire by repeatedly putting it in a Bunsen burner's flame for two or three seconds at a time and alternating this with dips into a 6 M HCl solution. This will clean any residue off of the wire so that interfering compounds are removed. Note the color of the clean probe in the flame. Dip the Nichrome wire into the tin salt solution. Put the end of the wire into the flame of a Bunsen burner. Observe the color of the flame above the wire *not the wire itself*. The color above the wire should be blue for this tin test but it may be difficult to see without some experience. Repeat this test until you have a good idea of the color of tin in the flame.

Perform the flame test on the tin sulfide solution that you boiled with HCl repeatedly and set aside. Repeat this flame test four or five times. Is there an interference that precludes the use of this test for tin? What color is the interference? If you have time use first a Nichrome wire then a glass rod as a flame probe and compare the results.

Separation of bismuth from copper and cadmium

Bismuth can be successfully separated from both copper and cadmium by adding an excess of ammonium hydroxide (NH_4OH). The supernatant after centrifugation will contain copper and cadmium and the precipitate will be $Bi(OH)_3$. Copper and cadmium can be detected in the supernatant by reducing copper(II) to copper metal with sodium hydrosulfite ($Na_2S_2O_4$) and then detecting cadmium as cadmium sulfide as before.

To the test tube labeled as insoluble Group II sulfides above add 2 mL of 6 molar HNO_3. Boil this solution for two minutes. Add another mL of nitric acid and boil for another two minutes. Centrifuge and collect the supernatant. Discard the precipitate in the appropriate container. This precipitate contains elemental sulfur (S_8) produced by the oxidation of S^{2-} and products from the thioacetamide reagent.

Basify the supernatant with 6 M ammonia solution. Add the base 1 mL at a time until the solution is basic to litmus (red litmus turns blue in base; blue litmus stays blue in base). After you are sure that the solution is basic add another 10 drops of ammonia solution. Bismuth will precipitate from this solution as $Bi(OH)_3$. Centrifuge, collect the supernatant for the copper test below, and collect the bismuth precipitate, wash it with two 2 mL aliquots of distilled water. This washed precipitate can be tested for bismuth by the addition of 5 drops of 5 M NaOH to the precipitate in a test tube and then the addition of 0.1 M $SnCl_2$ dropwise to the same test tube. The presence of bismuth will be confirmed by the formation of black metallic Bi^0.

The supernatant collected above will have the characteristic dark blue color of the copper amine complex—one of the most beautifully colored metal complexes that you will encounter. Confirm the presence of copper by adding 0.2 to 0.4 g of solid sodium hydrosulfite. The copper(II) ions will be reduced to copper(I) and (as you heat this test tube in a hot water bath) ultimately to metallic copper (black or coppery red). Centrifuge, decant, and collect the supernatant from this test. Discard the solid containing the reduced copper and any unreacted hydrosulfite in the appropriate container.

The supernatant from the last step will contain cadmium if everything goes as planned. To confirm the presence of cadmium, add 1 mL of 1 M thioacetamide and put the test tube in your hot water bath. Take the necessary precautions regarding the generation of H_2S. The formation of yellow cadmium should be forthcoming. Compare this precipitate to your CdS standard made from the pure $CdCl_2$ solution in your first procedure. Why should they be different?

Make sure that the solutions and precipitates that you have produced in your analyses are disposed of in the appropriate container. If you are not sure where this container is, ask someone who knows!

Group II unknowns

Get a test tube containing a Group II unknown from your instructor. Record the unknown number if this is applicable. Divide the solution into at least two portions so that you can repeat your separation scheme from the beginning if it becomes necessary. Assume that your unknown contains all the members of this group that we have studied. Write down the steps of the scheme as you perform them making careful notes about the success of each confirmation step. Repeat the known confirmation tests if your are allowed to and if the need arises because of an indistinct result. Report to your instructor your separation scheme and the Group II members present in your unknown.

Instrumental Analysis of Cadmium

Using atomic absorption spectrometry, the light absorption characteristics of mercury allowed us to sensitively and yet selectively detect mercury in complex samples (see the previous chapter). Cadmium can also be detected by AAS and this is often the instrumental analytical method used; however, another spectroscopic method is fast replacing AAS for the analysis of cadmium and many other elements, including most of the toxic metals. Inductively coupled plasma (ICP) also takes advantage of the special characteristics of atoms interacting with light; however, this time the phenomenon of atoms selectively <u>absorbing</u> light is not the key. Instead, the potential of highly excited atoms to <u>emit</u> specific wavelengths of light is drawn upon to create one of the most sensitive analytical methods yet devised. The detection limits for many elements using ICP—the smallest mass that can be detected for certain—is unsurpassed by any other analytical method. The lower the detection limit, the smaller the traces of toxics that can be detected and the smaller the samples that are necessary for an analysis.

The inductively coupled plasma

When atoms are heated to very high temperatures, the electrons in the atoms' shells are excited to higher electronic orbits (higher in an energy sense). After a very short time at this higher energy state the electrons return to a lower energy state by emitting a **photon**, a packet of light. Since, as we learned in the AAS section, the electronic states available to electrons are quantized (strictly limited) and thereby depend on what kind of atom (which element) is being excited, the wavelengths—colors or energies—of electronic emission are also strictly dependent on what kind of atom is being excited. This means that the light given off by a sample containing many different kinds of excited atoms will have the light "fingerprints" of the atoms in the sample. The colors of light emitted from an excited group of atoms will be characteristic of the atoms in the sample. Voilà! We have another method of sensitively determining what is in a sample: qualitative analysis. In fact, if we can successfully separate each of the different atomic light fingerprints from a light emitting sample we can detect many different elements simultaneously.

The heart of the ICP is a very, very hot **plasma** that can excite all of the atoms in a sample fed into it. This visually bright source is routinely referred to as the torch. The temperature in the torch is many thousands of degrees. The energy of the torch is powered via electrons that are stripped off of argon atoms and accelerated by a strong magnetic field; the heat of the torch is, in effect, *induced* by the *coupled* magnetic field accelerating the electrons in the plasma. Therefore, the name of this energy source is **inductively coupled plasma**.

The high energy in the torch destroys any chemical compounds or complexes (**matrices**) that exist in the sample as it excites the atoms. In other methods, these matrices might interfere with the analysis. Complex matrices that chemically bind metals are the downfall of the classical Qual scheme and a problem with AAS to a lesser degree also.

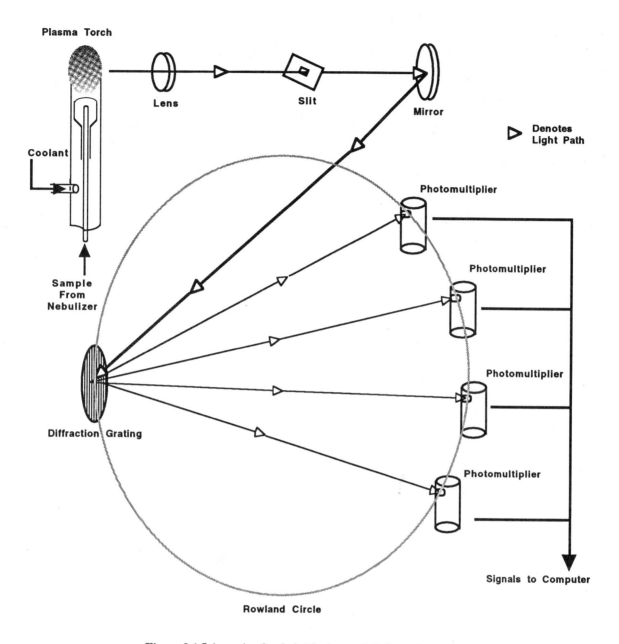

Figure 3.1 Schematic of an inductively coupled plasma instrument.

True multichannelling

After the sample is excited with the intensely hot torch, the light from the emitting atoms in the torch is collected via a set of lens and mirrors and focused onto an integral part of the ICP, the **diffraction grating**. The job of this grating can be compared to the monochromator's in AAS. The monochromator's job is to isolate a particular wavelength of light from the graphite furnace (or AAS flame) and focus that light on the photomultiplier tube (often called a PMT) while blocking all other wavelengths. Similarly, the diffraction grating in an ICP instrument separates in space each of the individual wavelengths emitted by the excited atoms and directs these individual beams of light onto many different photomultipliers positioned some distance away from the grating (0.5 to 3 meters). Each wavelength is focused on a different PMT, all at the same time. There is no need to analyze for each element sequentially as in AAS. The determination of fifty or more elements can be accomplished *simultaneously* in a few seconds by ICP. This is multichannelling. (There are ICP instruments that do not use multichannelling and instead scan from one element to another sequentially.)

A simple multichannel ICP

The schematic diagram on the previous page includes the major parts of an ICP instrument in simplified form: the torch, a lens to focus the emitted light, a slit to provide a rectangular image, and a mirror to reflect the emitted light onto the diffraction grating. This diffraction grating is also included in the diagram; and for clarity, just four PMTs instead of a possible fifty are shown. The position of the photomultipliers in relation to the grating approximates a circle because the grating separates the light from the torch into individual beams fanning out in a curve. This is called the Rowland circle.

A few more important parts of the ICP are not shown in this diagram. Most multichannel ICP instruments have slits just in front of each PMT to help keep light traveling to a nearby PMT from interfering. In addition, the entire Rowland circle—grating, slits, PMTs and all—are enclosed in a gas tight chamber that is flushed and cooled with an inert gas. This keeps the optics clean and keeps the heat from the nearby torch from affecting the temperature-sensitive PMTs.

Also not seen is the sampling device that feeds the sample into the torch. This is usually a **nebulizer** or electrothermal vaporizer. The job of a nebulizer is to suck up sample liquid from a sample vial and disperse it in tiny droplets into a stream of gas (usually an inert gas like argon) that is fed into the gas flowing into the base of the torch. Atomic absorption spectrometry can also use a similar device to introduce sample liquids into a acetylene/air flame instead of the graphite furnace that you were introduce to in the last chapter.

The job of an electrothermal vaporizer is to turn a solid sample into a gas that can be swept (again usually by argon) into the torch. This is similar to the AAS's graphite furnace described before and similarly allows for the analyses very small, yet environmentally important, solid phase samples like biological tissue or insoluble precipitates.

The calibration curve—a quantitative tool

One of the quantitative analytical tools that is central to both ICP and AAS (among many other analytical techniques) is the **calibration curve**. This is a means of determining just how an instrument responds to different amounts of an analyte, that is, the chemical you are interested in. This is necessary because no instrument responds in a perfectly theoretical way to all samples—a response that can be described by an instrument's manufacturer and used forever is unachievable. For any instrument, deviations from the theoretical are caused by any number of factors and must be taken into account. Since it is impossible to know all of the deviations exactly, calibration is a necessary part of any quantitative instrumental analysis.

This calibration relationship is usually represented as an x-y plot of analyte concentration versus instrument response. The concentration of the standards that are used to calibrate the instrumental response must be known as exactly as possible and therefore are often relatively expensive. In addition, to help correct for day to day drift of instrumental response, the calibration is often performed the same day that the samples are analyzed instead of far in advance or long after sample analyses.

In our study of the ICP analysis of toxic cadmium, the calibration curve is the relationship between the ICP's response (intensity) versus different masses of cadmium introduced into the instrument one at a time.

For instance, here is a series of 5 standards (samples with known Cd concentrations) and the instrumental response to each one (in arbitrary units of intensity taken by the computer from the PMT output):

Cadmium mass present in standards	Intensity
1 pg	15
3 pg	44
10 pg	155
25 pg	373
50 pg	751

An x-y plot of this relationship appears below.

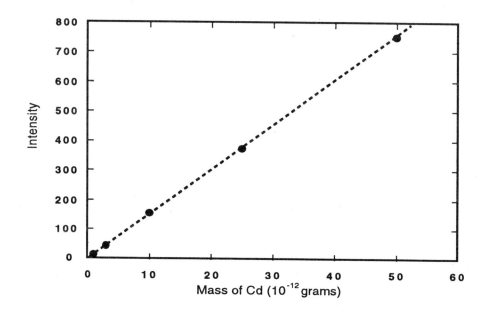

Figure 3.2 Cadmium ICP calibration curve.

As you can see, the response of the instrument (PMT output) over the range of masses analyzed is almost completely linear. This means, for instance, that doubling the mass of cadmium in the sample doubles the instrument's response to that sample. Although this plot does not show it, the response of the instrument at higher cadmium concentrations was not linear, and samples falling outside of the linear range have to be diluted before being analyzed. The resulting cadmium concentration is them multiplied by the dilution factor to produce the true cadmium concentration.

A calibration plot is usually "fitted" with a line drawn through the points to help in the use of the plot for future analyses. An analyst who has performed this calibration now knows that the concentration of future samples (with masses in this range) can be accurately determined simply by measuring the instrument's responses to the samples and referring to the calibration curve (line). In practice, this relationship is stored by the instrument's computer and the report that the analyst generates from the computer will detail, for instance, simply a sample's identity and concentration instead of the instrument's response.

ICP EXPERIMENT—
DETERMINATION OF CADMIUM

The inductively coupled plasma determination of cadmium is based on the detection of the emission of cadmium in the high temperature plasma. As with the atomic absorption experiment briefly described at the end of the previous chapter, another less toxic element can be used to avoid additional cadmium waste generation. A dissolved iron solution is a good candidate.

1. Prepare the ICP instrument following the manufacturer's instructions. The computer based nature of this process makes this procedure very specific depending on the instrument used, and it will be performed by a skilled operator. The better the computer interface the less previous knowledge the operator must have of computer systems and the better the reproducibility and accuracy of the analyses. In addition, the safety aspects of the instrument are automatically handled by the computer. This is exactly what computers were meant to do.

2. One of the routine procedures performed daily with most spectroscopic methods and with ICP is calibration using standards of known concentration. The instruments on the market now (1993) will accept multiple standards (50 or more) of multiple different elements in almost any order (high, low, middle concentration, etc.). The computer then plots the resulting calibration curve for future output and use by the computer in subsequent analysis of unknown solutions. If elemental standards are available watch the operator perform the calibration and examine a printed output of the results.

3. To examine the precision of this technique, use one of the calibration standards as an unknown solution and get a printout of the computer's calculation of that sample's concentration. A routine precision of < 5% is easily accomplished with this instrument.

Chapter Three Problems

1. If cadmium is so toxic, why is it used to electroplate metal parts?

2. Why is cadmium almost always present to some extent in products made or coated with zinc.

* 3. The mud at the bottom of New York Harbor and the nearby Hudson River is extensively contaminated with cadmium. Why?

4. What physical characteristic of cadmium makes it useful as a semiconductor?

* 5. What have you done with NiCd batteries that you removed from a watch or calculator or toy? Why? What effect will your action have on the toxic cadmium in our environment?

6. What happens to toxic cadmium in the particulate matter from a burning cigarette that is not filtered out in the smoker's lungs?

7. What organs store ingested cadmium? Why are these organs the site for storage of this toxic metal?

8. What apparent relationship do zinc and cadmium have in the body?

9. Have scientists who have tried to induce high blood pressure in laboratory animals been successful by feeding them cadmium containing diets?

10. Define the word chelation and describe how a chelation agent is used.

11. How could you sample a smoker's household for particulate cadmium?

12. Thioacetamide, CH_3CSNH_2, reacts with water to produce H_2S and acetamide, CH_3CONH_2. Write the equation of this reaction down and balance it.

13. Write down the net ionic equation for the reaction of cadmium chloride and hydrogen sulfide that is used in the first procedure of the Qual scheme.

14. Why is it a good idea to stopper the product of this reaction?

15. Find and write down the solubility products for all of the sulfides of the metals that you analyzed in this chapter.

16. Why is washing precipitates an important step in qualitative analysis?

* 17. Can you think of any reason why your standard tin solution might give a different flame test compared to the tin solution that you separated and flame tested?

* 18. What happens to the sulfide ion present in the tin sulfide solution as you boil this solution with HCl.

19. What is the quickest and easiest way to detect Cu^{2+} ions in solution?

* 20. Write down and balance the oxidation/reduction reaction that is used to confirm the presence of bismuth. Show only the ions that are oxidized and reduced (with their oxidation states).

21. Make a flow chart of the Group II Qual scheme analyses for all the metals that your separated in this chapter?

* 22. What happens to the chemicals from these procedures if you poured them down the drain?

23. Of the two instrumental methods studied so far, one involves the emission of light and the other the absorption of light. Which is which?

* 24. As far as environmental analysis (or quantitative analysis in general) is concerned, is it better to have higher detection limits or lower detection limits? Why?

25. How does ICP get its name?

26. Why are chemical matrices less of a problem in ICP than in AAS or the classical Qual scheme?

27. What is the difference between the job of the monochromator used in AAS and the job of the diffraction grating used by ICP?

28. How is more than one element detected at a time with ICP? Why is this impossible with AAS?

29. What is the Rowland circle?

30. How many photomultiplier tubes are necessary for an ICP instrument?

31. What are two sampling devices used by ICP instruments and how are the samples that they handle different?

32. What are the variables plotted in a typical calibration curve?

* 33. Why is a linear calibration plot called a calibration curve if it is best described as a line?

34. Why are analytical methods calibrated?

35. What is the purpose of the calibration curve?

*36. How do you know when an analytical instrument is not responding in a linear way to different concentrations of sample?

*37. How can an astute analyst handle the situation in question 34?

*38. Using the calibration curve in Figure 3.2 estimate the concentration of cadmium in an environmental sample that exhibited a light intensity of 500 units.

*39. Using the calibration curve in Figure 3.2 calculate the equation for the line displayed using two sets of points on the line (50, 751) and (10, 155). Assume that the line passes through the origin.

*40. Using the equation for the line calculated in problem 39 calculate the concentration of cadmium in an environmental sample that exhibited a light intensity of 500. units. How does this answer compare to your estimate?

Chapter 4

Chromium: Toxic or Essential?

VITAMINS AND MINERALS

When is a toxic metal not toxic? This question had little meaning before this century because the knowledge of what was actually required for human health was not very well understood. For humans, the dietary requirements of organic compounds that are not actually food (called vitamins) were discovered slowly over the last two hundred years. Evidence that the absence of specific substances in human's diet caused disease was discovered a long time ago: A Scottish doctor determined in the 18th century that sailors developed a disease called scurvy when they were at sea for long periods of time without fresh fruit and therefore deprived of its vitamin C. The British navy solved this problem by giving its sailors limes on their voyages to supply their daily Vitamin C dosage. This process earned British sailors the moniker "limeys."

In addition to the evidence that specific organic molecules like ascorbic acid (Vitamin C), retinol (Vitamin A), and alpha-tocopherol (Vitamin E) were necessary for good health, it was also discovered that there were trace minerals or metals that were needed too. The list started off rather small and included metals that were not generally considered toxic such as iron, manganese, copper, and zinc. Later it became clear that even metals that were known to be toxic to humans were required in the diet in very small (trace) amounts. These metals, known to be required in a healthy diet, include selenium, arsenic, and chromium. These three are all also known to be toxic to humans at concentrations somewhat above that required for nutrition. The small levels at which they are needed has earned these nutrients the title ultra trace elements.

Humans that have diets missing these metallic elements over long periods of time are more susceptible to diseases, bacteria, viruses, and the toxic effects of other chemicals. The answer to why this is true is actually unknown but probably stems from the ability of our bodies to adapt to our chemical environment. Since these toxic metals are present in minute amounts in the natural environment, human beings have always been in contact with them through our diets, water, and contact with soil and dust. Because the level of these species was low enough not to cause damage, our bodies learned to make use of them in organometallic complexes. Hemoglobin—with a single iron ion as an integral part—is an example of a very important, large, organometallic complex required for life in animals. The role of many chemical species are not understood very well even though they are known to be important in trace amounts. The answer to the question "when is a toxic metal not toxic?" is when the metal is essential in minute amounts for a healthy diet.

Chromium Sources in the Environment

The major sources of exposure of humans to chromium (other than the natural background present in food—about 60 µg per day) are predominantly in industries that manufacture chromium contain chemicals or those involved in electroplating where work sites are contaminated with chromium salts or chromium carried into the air by aerosol sprays. Interesting also is the possibility of chromium ingestion of stainless steel welders who actually vaporize chromium containing alloys during the welding process.

Finally, as with many other metals like aluminum, cadmium, lead, and nickel, chromium is also released in second hand cigarette smoke that can be inhaled by people other than the smoker. Since many of these metals have been associated with cancer in animals, their contribution to the carcinogenic abilities of tobacco smoke is a subject of interest. The United States government's recent recognition of the dangers of second hand smoke should help to emphasize the importance of looking at more of its components.

Chromium Toxicity

Chromium exists in chemical forms other than as the pure metal that you think of on old automobile fenders. In the environment, chromium forms two common oxidation states, hexavalent chromium, Cr^{6+} and trivalent chromium, Cr^{3+}. These two species form different compounds and subsequently exhibit different levels of toxicity for humans. Hexavalent chromium forms more water soluble compounds than the trivalent form. The link between inhaled chromate (Cr^{6+}) salts and lung cancer in the chromate and pigment industry has been made by research stretching back to the 1950s. Workers in these industries are also be more likely to have higher chromium concentrations in their livers and kidneys. Their bodies work to eliminate these compounds by excreting them in the urine. Most clearly demonstrable is major damage to skin and mucous membranes that come into contact with hexavalent chromium compounds. The name for the disease **chromate eczema** was coined for workers who have a skin inflammation after coming into contact with chromium containing cement. These same workers sometimes also had painless skin ulcers called chrome "holes" in the skin on the back of their hands from prolonged cement contact and perforations in the septum in their noses from breathing cement dust.

In regards to the cancer causing abilities of chromium, it is generally accepted that Cr^{6+} is more toxic than chromium(III) but this is actually probably a solubility effect rather than the way that the actual damage occurs in cells. Chromium(VI) forms chemical compounds that will readily cross through the cell wall and thereby have easy access to cellular material inside the body (and inside the cell). Originally it was thought that after entering these biological systems Cr^{6+} was apparently reduced to Cr^{3+} inside the cell itself (*in vivo*), and it was this trivalent chromium that binds and distorts the genetic material in the cell, subsequent causing mutation and cancer. Recent research on the dangers of chromium has centered on detecting a Cr^{5+} intermediate complex created by the reduction of Cr^{6+} *in vivo*. The creation of this intermediate may in turn promote the production of very reactive chemical species inside cells called **radicals** that in turn break strands of **DNA**, disrupt the flow of genetic information, and cause mutations and cancer.

Sampling for Chromium in the Environment

The detection of airborne chromium dust is an important aspect of controlling the damage of this toxic element. In industrial plants that produce chromate chemicals and pigments the level of chromium containing dust is carefully monitored and controlled in order to protect the lungs and nasal passages of the workers. Sampling procedures for chromium containing dust is similar to that of cadmium: air that contains dust is pulled (by a pump) through a filter with small pores to catch the dust particles. If the rate of flow (for instances, liters of air per minute) is known, then a specific volume of air can be sampled by controlling the time the pump runs. After sampling a known volume of air through the filter, the filter is washed with an acid or chelating agent and the resulting solution can be analyzed for chromium.

Tests that examine the chromium level in the blood of workers before and then after their work shifts can be used to assessed blood levels created (absorbed by the body) by this occupation. These analyses are actually very difficult to accomplish because the stainless steel needles used in drawing blood can contaminate the sample with traces of chromium and give erroneous results. This is especially important for monitoring the blood's chromium level in people with very low amounts. Careful sample handling and the establishment of the background chromium level is very important in these procedures.

Qualitative Analysis of Chromium and Group III Cations

The Group III cations number eight in the classical Qual scheme. The complex mixture of all the members is initially precipitated as sulfides which can then separated into two subclasses by acid addition. We will involve ourselves with only four cations in Group III, two from one subclass and two from the other. Our focus will be on chromium, nickel, cobalt, and iron. To begin, we will precipitate all four members in the mixture using thioacetamide as you did in the Group II scheme of the last chapter. Again, take extra precautions with these steps because of the danger of hydrogen sulfide. A hood is mandatory! While this procedure has its drawbacks, it is still considered a better alternative that constructing a hydrogen sulfide generator and "piping" the gaseous product into your reaction vessels. In addition, the relatively slow production of H_2S by thioacetamide makes it easier to control than the gas generator.

The chromium confirmation test

After separating chromium from the others members of the group (as chromate, anion) you will detected its presence by converting chromate to dichromate and then oxidizing dichromate to chromium peroxide. Visually detecting the distinctive (yet transitory) blue color of chromium peroxide is a confirmation of the presence of chromium. Let's practice that test first.

Put five drops of a $0.1\ M$ solution of potassium chromate in a clean test tube. Note the color of this solution. Add 2 drops of $6\ M$ HCl. Note the color change. The new color is the color of the dichromate anion. Oxidize your solution of dichromate to chromium peroxide with the dropwise addition of 3 % hydrogen peroxide (H_2O_2). Only a few drops should be necessary to see the blue confirming color of chromium peroxide. If you set this test tube aside in your test tube rack you will see this color soon fade away.

Precipitating the Group III cations

Put a medium-size beaker two thirds full of tap water on a hot plate and bring it to a boil. This is your hot water bath.

To approximately 3 mL of a solution containing the soluble salts of the four selected members of this group add enough $6\ M$ NH_3 to make the solution basic to litmus. After the successful litmus test, add an additional 0.5 mL of $6\ M$ ammonia solution. To this basic solution add 1 mL of $1\ M$ thioacetamide solution and put the test tube in your water bath for five minutes under the hood. After five minutes stir the precipitate and then heat for three more minutes.

Separate the supernatant from your Group III precipitates by centrifuging for one minute. Decant the supernatant and wash the solid with first 2 mL of $2\ M$ NH_4Cl solution (with the necessary stirring, centrifuging, and subsequent decantation) and then 2 mL of distilled water. Discard the supernatant and washes in the appropriate container.

This washed solid contains the sulfides and hydroxides of the four members of this group. Two of the members, cobalt and nickel, can be separated by acidifying this solid mixture and separating the more soluble sulfides. These steps are next.

Separating and confirming the presence of Co and Ni

Acidify the washed precipitate from the last step with 2 mL of $3\ M$ HCl; stir the mixture and transfer all of this material into a small beaker. Put the beaker on a hot plate and boil: watch for splattering. Use a watch glass to cover the beaker if necessary.

After about a minute of boiling add 1 mL of distilled water and pour the entire mixture into a test tube using sparingly small amounts of water from your wash bottle to make the transfer complete. Centrifuge the tube and decant the supernatant into another test tube clearly marked. This tube contains the acid soluble sulfides of chromium and iron so a label like "Cr + Fe" would be good.

Wash the precipitate collected above with first a volume of 6 M HCl approximately equal to the volume of the solid itself and then follow this with a double volume of distilled water. Spin down each wash and discard the liquid above the precipitate into an appropriate container. Dissolve the washed precipitate in 2 mL of dilute aqua regia (HCl + HNO_3) by adding 1 mL of 6 M HCl and 1 mL of 2 M HNO_3 to the precipitate in the test tube and putting the test tube in boiling water until dissolution is complete. Add 2 mL of distilled water to the tube after complete dissolution. The next step is to confirm the presence of first cobalt and then nickel.

Divide the solution in the test tube in half and add (dropwise) about 1 mL of 0.3 M potassium thiocyanate solution (dissolved in alcohol or amyl alcohol and ether). The presence of cobalt will be confirmed by the formation of a blue solution of cobalt thiocyanate anion, $Co(SCN)_4^{2-}$.

To the second half of the solution divided above add dropwise a 1 % alcohol solution of dimethylglyoxime ($C_4H_8N_2O_2$). The confirmation for the presence of nickel is the formation of a rose-red precipitate.

Dispose of the material left over after your tests in the appropriate containers. Ask your lab instructor if you do not know where or what the correct container is.

The separation and detection of iron and chromium

To the "Cr + Fe" labeled solution collected above add 8 M sodium hydroxide (NaOH) solution dropwise until the solution becomes basic to litmus. Add 0.5 mL of additional base to the test tube. Pour the test tube contents into a small beaker, add 5 drops of 3 % H_2O_2 slowly, stirring between each drop. Wait one minute after the last drop addition and then heat the beaker to a boil for one minute. Add a few mL of distilled water to keep from boiling the beaker dry. Quantitatively transfer the contents of the beaker to a test tube using small squirts from your wash bottle if necessary. Add water until the final volume is five mL. Centrifuge the test tube and decant the supernatant into a clearly labeled ("Chrome ?") test tube. The precipitate should contain iron hydroxide and the separated liquid should contain chromate anion.

Test the precipitate for iron by adding 2 mL of 2 M sulfuric acid (H_2SO_4). Heat the solution in the test tube in the water bath until the solid completely dissolves. To this solution add 1 or 2 drops of 0.3 M potassium thiocyanate solution. The presence of iron will be confirmed by the deep red color of the iron thiocyanate complex.

The supernatant solution decanted above and labeled "Chrome ?" is basically the same solution that you started with at the beginning of the procedure in "The Chromium Confirmation Test." The mixture of the original four elements has been winnowed down into a final solution containing just chromate. Perform the identical steps as before, transforming chromate into orange dichromate with acid, and oxidizing dichromate to chromium peroxide (blue). If you will look back now at the previous example of this test stored in your test tube rack you should see that the original blue has faded away. Why?

Dispose of the material left over after your tests in the appropriate containers. Ask your lab instructor if you do not know where or what the correct container is.

Group III unknowns

Get a test tube containing a Group III unknown from your instructor. Record the unknown number if this is applicable. Divide the solution into at least two portions so that you can repeat your separation scheme from the beginning if it becomes necessary. Assume that your unknown contains all the members of this group that we have studied. Write down the steps of the scheme as you perform them making careful notes about the success of each confirmation step. Repeat the known confirmation tests if your are allowed to and if the need arises because of an indistinct result. Report to your instructor your separation scheme and the Group III members present in your unknown.

Instrumental Analysis of Chromium Using Ultraviolet/Visible Spectroscopy

The power of analytical spectroscopy is its use as both a qualitative *and* quantitative tool to determine not only what species are present in a sample but also how much. In the environmental sampling of chromium in the blood, for instance, the mere detection of the presence of chromium in a steel worker's blood is not all that is required. For a blood test for mercury this might be adequate to determine that the worker has had some exposure to toxic mercury and the source should be determined and eliminated. (A quantitative test would also not doubt be performed to determine the extent of the exposure.) But for the determination of chromium in blood, the <u>amount</u> of chromium present (per deciliter of blood, for instance) is very important because some chromium will naturally be present and is, as we learned before, nutritionally required. Therefore, the analysis must be both qualitative and quantitative.

Atomic spectra

Our introduction to atomic absorption and inductively coupled plasma spectroscopies in earlier chapters described, first, an analytical method of elemental analysis based upon the light *absorbed* by atoms in the gas phase (AAS). The sample is introduced, by nebulization of a solution of the dissolved metal, into a air/acetylene flame and a source light beam where they absorb some of the light beam. The sample introduction system in ICP is substantially the same as in AAS; however, in ICP the atoms in a very hot plasma (not a flame) are detected not by their absorption of specific wavelengths of light but by their *emission* of specific wavelengths.

The emission and absorption spectra of atoms are relatively simple because there are only a few energy states available for occupation by the electrons being promoted to higher energy states (in absorption), and few lower energy states to drop back to from higher states (in emission). Remember that the energy states are quantized and thereby limited in number. This limitation of electronic places to move to or from means that there are only a few atomic electronic transitions that actually occur, a few absorption or emission lines that can be detected, and therefore a simpler spectrum showing the lines.

Molecular spectra

Molecular spectra are somewhat more complex than atomic spectra because the number of possible electronic transitions is much greater. Molecules are much more complex in their electronic structure than single atoms and therefore provide many more possible transitions than single atoms. Molecular absorption spectra—showing the light missing as it exits a sample containing light absorbing molecules—reflect this complexity. Similarly, molecular emission spectra—showing the additional light exiting a sample containing light emitting molecules—contain the fingerprints of many different transitions. Both absorption and emission spectra of molecules can therefore be more complex than atomic spectra.

Figure 4.1 is a comparison of an atomic spectrum and a molecular spectrum that contains the same metal element, chromium. Both plots show wavelength on the x-axis and absorption on the y-axis. The upper spectrum is the gas phase absorption spectrum of <u>atomic</u> chromium. Notice the sharp, well define structure of the absorption lines. One of these sharp lines would be used in AAS to detect and determine the concentration of chromium in a sample. The lower spectrum is of a molecular complex containing chromium in aqueous solution, chromium diphenylcarbazide. This spectrum shows the additional broadening and complexity of the chromium absorption in a molecule as opposed to an atom. The absorption features have changed from individual lines to a much broader feature called a band. This broadening effect is actually the result of a number of factors that we won't cover here but include the resolution of the detectors used in molecular spectroscopy, the intensity of the lines themselves, the fact that the molecule is in a solution as opposed to the gas phase, and the temperature of the samples; atomic spectra are often taken at high temperatures and molecular spectra in UV/vis spectroscopy are mostly at room temperature. The result is that molecular spectra are noticeably different from atomic spectra.

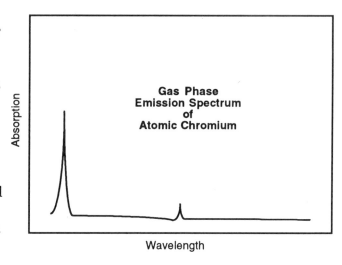

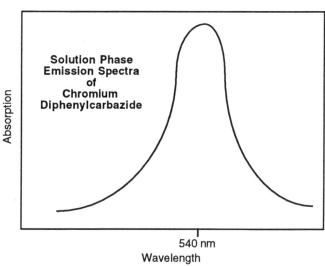

Figure 4.1 Comparison of atomic and molecular spectra.

Ultraviolet and visible wavelengths

The wavelength region of the electromagnetic spectrum that humans can see is approximately 380 to 780 nanometers (380 x 10^{-9} to 780 x 10^{-9} meters). As we learned earlier, wavelengths of light correspond to the colors that we see with our eyes: dark blue or violet light is what 400 nm light looks like; 550 nm light appears green, and 700 nm light looks dark red. Light of longer or shorter wavelengths than the 380 to 780 nm extremes is invisible to us because the chemical reactions in our eyes don't respond to these longer or shorter wavelengths. (Some other animals do in fact have a different color range than us, and some are completely color blind.) The region of higher energy light above violet is called the ultraviolet (*beyond* violet); the lower energy light below red is called infrared. The light in both the ultraviolet (UV) and infrared (IR) regions and all the light beyond them is invisible to human beings.

The energies of the light waves interact with matter in different ways. Molecular spectroscopy of the nature that we are interested in—involving molecules containing metal atoms like chromium—uses light in the UV and visible wavelength regions because that is where these molecules interact best with light. The light that is shined through samples in spectroscopy is chosen careful so that the interaction of light with the species in the sample is well known. If the analyst chooses the wrong wavelength of light to probe the presence of, for instance, a toxic metal complex in an environmental sample, then the toxic species would not absorb at the chosen wavelength, would be invisible, and would not be detected. This could be disastrous.

The two previously introduced spectroscopic techniques, AAS and ICP, used the relationship between light and matter as a tool to detect the presence of trace species in environmental samples. They also allow the analyst to determine the concentration of these chemicals at vanishingly small levels. The instrumental determination of chromium as it is described below is still more sensitive than the normal Qual scheme yet is less sensitive in its detection limit for chromium than either AAS or ICP. The historical and practical significance of ultraviolet/visible spectrophotometric determination of chromium, however, warrants its inclusion as a worthy instrumental means of determining this toxic metal. In many industrial settings, steel plants, the pigment industry, and electroplating shops this method is still widely used because of its simplicity, speed, and low cost as compared to the other instrumental methods.

A simple double beam UV/vis spectrometer

The analytical instruments that we examined in the previous chapters are actually more complicated that the normal ultraviolet/visible spectrometer; however, some of the same instrumental parts that we were introduced to in AAS and ICP are still present here. Like the atomic absorption instrument, the UV/vis spectrometer has a source of light, usually two different lamps: one that produces strong ultraviolet wavelengths and a separate lamp for visible wavelengths. (See Figure 4.2.) These lamps are not light sources as in AAS—emitting to a particular wavelength, specific to a certain elemental line. Instead they are broad band sources that shine light over a wide wavelength range. During the UV/vis spectrum scan, the lamps beams are automatically switched at the appropriate wavelength (time) so that for all practical purposes the sample is exposed to a continuous scan of light from the far UV through the visible. With some instruments, if you listen carefully during the scan you can hear the motor that rotates the source selecting mirror as it switches from one lamp position to the other.

The first instrumental component that the source's light "sees" after leaving the source lamp is a wavelength selector or monochromator. Similar to that described in the AAS chapter, this device cuts the broad band of light coming from the source lamp down to only a single or very few wavelengths. The wavelength selector can be set a one specific wavelength as we will see below or scanned through a range of wavelengths to yield an entire spectrum. As the scan is performed from, say, 300 to 650 nm, the sample in the sample cell (see below) is exposed to (and gets the chance to absorb) light of first 300 then 301 then 302 nm and so forth. This continues until the scan is completed at 650 nm. The instrument keeps careful track of what wavelength is being passed through the sample and plots out that information on the spectrum.

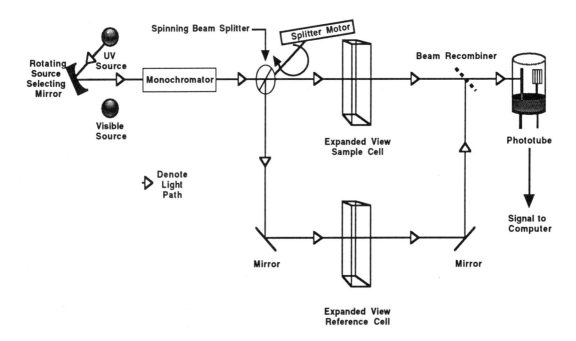

Figure 4.2 Schematic diagram of a double beam spectrophotometer.

The next instrumental part in the UV/vis spectrometer is the sample cell. This *is* different from both AAS and ICP since those methods use gas phase sample "cells." In this spectrometer, the sample (in our environmental example) is a metal complex dissolved in a solvent. The samples for this instrument are always liquid. A small volume of the liquid sample is placed in a special glass contained called a **cuvette**. The cuvette is square in two dimensions and the glass faces, through which the light passes, are designed to be as flat as possible so that no light is reflected and wasted or allowed to cause interference. The cuvette's glass is made of a substance that does not absorb in the UV or visible wavelengths. Finally the path length that the light travels inside the cuvette and therefore inside the sample is known accurately. Figure 4.2 has an expanded view of the sample cuvette. The actual path length of the sample cell is usually just one centimeter.

The last part of this UV/vis spectrometer is another instrumental component common to both AAS and ICP. It is a phototube or light detector. The job of the detector is to determine the amount of light of a particular wavelength that has successfully pass through the sample from the source. The instrument collects this information as an electric current produced by the phototube: more light means more current from the detector.

It is important to note here that the result of this scan is an x-y plot of detector output (y-axis) versus wavelength (x-axis). The data can be displayed as a plot of the percentage of light that successfully *passes through* the sample solution versus the wavelength at which this transmission occurs. This is a **transmission spectrum**. Alternately, the data can be displayed as a plot of a function of the light that was absorbed by the sample versus the wavelength, and this is called an **absorption spectrum** as in Figure 4.1. These two different ways of displaying the spectrum appear as the inverse of each other. Figure 4.3 below is the absorption and transmission spectrum of a compound plotted simultaneously over the same wavelength region on the same graph.

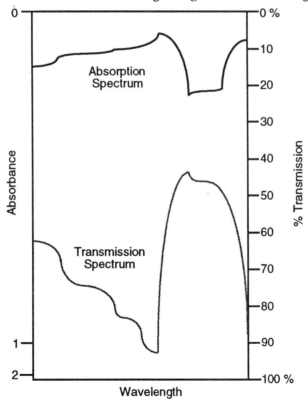

Figure 4.3 Double plot of the absorption and transmission spectra of same compound.

While all of the data plotted in Figure 4.3 use the wavelength axis (x-axis), the darker absorption trace is plotted in relation to the left vertical axis (left y-axis), and the transmission data uses the right vertical axis. Look carefully at these spectra and you will see that as transmission increases to higher and higher percentages, the absorbance decreases conversely. As absorption increases, less light is transmitted through the sample. Also notice that the features of the % transmission (or transmittance) spectrum and the features of the absorbance spectrum mimic each other but by different amounts (or different intensities). This is because of the way that the absorbance function differs from % transmission. Though both means of plotting these data are used by **spectroscopists**, the wavelength versus absorption plot is used more often than transmission.

The missing component of our double beam spectrometer is actually the part that makes it double beam. The sample cell containing the sample solution is designed to be invisible (completely transparent) in the UV and visible wavelengths. Actually it isn't. The solvent in which the sample species are dissolved is also chosen to be completely non-absorbing. In reality this is not achievable either. And finally there are other parts of the instrument that adversely affect the data obtained such as the drift of the light source and the spectral response of the detector. Taken together these problems could add up to make the job of getting an accurate spectrum of the chemical species of interest (the analyte) very difficult. There is an answer. The solution is to use a **reference beam**, the second beam in our double beam instrument. A reference beam is created by splitting half of the light beam after it leaves the monochromator but before it enters the sample. This beam, 50 % of the monochromator's output, is obtained by using a rotating circular disk that has a mirror on only one half. As the motor powered disk spins, the light shinning on it from the monochromator is alternately sent either through the sample cuvette or to another cuvette called the reference cell. (The reference beam is often directed around inside the instrument by highly polished mirrors as seen in Figure 4.2.) After passing through the reference cell the reference beam is recombined with the sample beam (recombined in space, not in time) and directed on to the phototube for detection. The reference cell contains everything the sample cell contains except the analyte molecules. In the case of our chromium diphenylcarbazide dissolved in water, the reference cell would contain a solution with everything but the chromium atoms. In this way the reference beam "sees" all that the sample beams "sees" except the analyte. The two beams (reference and sample) arrive at alternate times at the detector and the computer running the spectrometer simply subtracts the reference beam's signal from the sample beam's signal. Voilá: the signal generated is only the direct result of the analyte's signal, devoid of the complicating factors of cell and solvent absorption and source and detector drift. The benefits of a double beam spectrometer are great: quicker warm-up time because you don't have to wait as long for the phototube and source to stabilize and cheaper parts that still produce quality spectra.

Determining chromium in an environmental sample

The spectrum of the chromium complex that we are using, chromium diphenylcarbazide, is well known. The job of our spectrophotometer is not to reproduce that spectrum over and over again. Instead, we use the relationship between chromium concentration and the amount of light this complex absorbs to find out how much chromium is present: at the peak of the chromium complex's absorption, about 540 nm (see the bottom of Figure 4.1), the amount of chromium in solution will bear directly on the amount of light absorbed at that wavelength. Given excess diphenylcarbazide, zero chromium concentration will show little or no light absorption at 540 nm. (The reference blank should take care of any of this "nonchromium" 540 nm absorption—see below.) Some chromium in solution will show some absorption; more chromium means more absorption, etc. This relationship is sensitive enough to detect small amounts of chromium and linear over a wide enough range to make it analytically useful.

Chromium can be determined in an environmental sample (such as paint chips that might contain a chromium pigment) by oxidizing all of the contained chromium to Cr^{6+} with a strong oxidizing reagent and mixing with sulfuric acid—a process called digestion. After completely dissolving the sample or after extensive digestion followed by filtering, the pH of the sample solution is adjusted to a specific value and the diphenycarbazide reagent added. The characteristic color of the solution develops over a few minutes time and the absorption of the solution is measured with the spectrometer "tuned" to 540 nm. [Given the chromium complex's absorption spectrum in Figure 4.1, what do you this the color of this solution is?] The reference cuvette is filled with all of the same reagents except the dissolved paint chips and is placed in the reference beam to allow the instrument to account for all of the factors that might cause light absorption not due to the presence of the chromium complex itself. Distilled water can be taken through the same steps as the paint—oxidation, acidification, and reagent addition—so that the reference will contain everything except chromium.

Next a calibration curve of chromium concentration versus absorption is created using the same chemical matrix and known amounts of chromium. A single calibration curve can be used day to day for many different digestions but usually a calibration curve is run the same day as the samples in an effort to cancel out any day to day drift of lab conditions, water contamination, changes in procedure, etc. These precautions might not be necessary if the range of the chromium content is well known and varies very little from day to day as in a chromium electroplating bath.

If the sample's chromium concentration falls somewhere above the linear range of the calibration curve (see Chapter 3), the concentration is therefore not know with any real certainty and less paint is used in a repeat digestion and the process repeated until the sample's absorption falls inside the linear range. Alternately the sample solution can simply be diluted with the chemical matrix and the pH adjusted and rerun. If the chromium content falls below the linear range or is undetectable then more paint chips must be digested in hopes of determining the small amount present.

UV/VIS EXPERIMENT—
DETERMINATION OF DICHROMATE

Acidic solutions of potassium dichromate are colored, and the intensity of this color is proportional to the concentration of the dichromate anion in solution (over a relatively wide range). The creation of a calibration curve and the subsequent determination of the dichromate concentration in an unknown solution is a simple experimental way to be introduced to ultraviolet/visible spectroscopy. This also does not involve any additional chromophoric reagents such as diphenylcarbazide as describe before. This procedure is, however, less sensitive than the procedure using diphenylcarbazide reagent.

The following experiment is written in a way that is generally applicable to both single beam and double beam spectrometers.

1. Turn on the spectrometer and let it warm up.

2. Dissolve 0.282 grams of potassium dichromate ($K_2Cr_2O_7$) in water in a 100 mL volumetric flask and dilute to the etched mark on the neck. Put the top on the flask and invert several times to mix the solution. This stock solution is a 1000 ppm chromium solution.

3. Pipet 2.00 mL of the stock solution into a 100 mL volumetric flask.

Add 3 mL of 9 M sulfuric acid solution (50 % vol/vol) to the flask and swirl gently. Dilute with water to the etched mark, and then invert repeatedly with the top in place to mix the solution. This is a 20 ppm Cr solution.

4. Repeat this dilution placing 5.00 mL of the stock solution and then 10.00 mL in separate 100 mL volumetrics. Remember to add 3 mL of 9 M H_2SO_4 solution to each volumetric before diluting to the line in each with water. These two dilutions are 50 and 100 ppm in chromium respectively. These three solutions (20, 50 and 100 ppm) are the standards that will be used to generate the calibration curve.

5. Make a blank solution by adding 3 mL of 9 M H_2SO_4 solution to a 100 mL volumetric flask, diluting to the line with water, and inverting.

6. Set the instrument's wavelength to 440 nm.

7. For a single beam instrument (and following the manufacturer's instructions): Put the blank solution in a pre-cleaned cuvette, place it in the sample holder, and adjust the absorbance to read 0. Next, set the absorbance to ∞ with the shutter in place.

For a double beam instrument (and following the manufacturer's instructions) Insert the blank cuvette and zero the instrument or store the blank's absorbance for subsequent subtraction. Some instruments require that you leave the blank cuvette in place during the standard and sample readings that follow.

8. Starting with the 20 ppm solution, measure the absorbance of each standard by rinsing out a pre-cleaned cuvette with a small amount of the standard solution and then filling the cuvette with that standard. Put the cuvette containing the standard in the sample chamber, close the cover, and read and record the absorbance.
Repeat this procedure with each of the standards, rinsing the cuvette each time with a small amount of the new solution before filling it for the reading. You will have three absorbance readings when you finish, each corresponding to a different chromium standard's concentration.

9. Rinse the cuvette with first deionized water and then a small amount of the unknown solution to be determined. As before, fill the cuvette with the unknown chromium solution and record the absorbance. If the absorbance falls above that of your most concentrated standard (larger absorbance) then ask your instructor how to perform a 50 % dilution of the unknown solution keeping the acid content roughly similar to the standards. Determine the absorbance of the diluted solution.

10. Make a plot of absorbance (y-axis) versus chromium concentration (x-axis) for the three standards.

11. Using the absorbance of the unknown solution that you recorded in step # 9, determine the unknown's concentration from the standard calibration curve. Don't forget to take into account any dilution you may have performed on the unknown: if you performed a 50% dilution then multiply the chromium concentration (extrapolated from the standard calibration curve) by 2.

12. Make sure that the solutions that you have produced in your analyses are disposed of in the appropriate container. *If you are not sure where this container is, ask someone who knows!*

CHAPTER FOUR
PROBLEMS

1. What is the difference between a vitamin and a mineral?

2. List at least two vitamins. Have you heard about any toxicity associated with these vitamins?

3. List at least three minerals (metals) that are toxic in large doses.

4. Where did the term "Limeys" come from as used to refer to British sailors?

5. How can it be that a toxic metal is also required by our bodies for good health?

6. What is your most significant source of chromium?

7. If Cr(III) does most of the damage in cells, why is Cr(VI) considered more toxic?

8. If a steel welder has **higher** levels of chromium in her blood after a work shift than before, what does this mean?

9. If a steel welder has **lower** levels of chromium in her blood after a work shift than before, what does this mean?

10. What is a major sources of contamination for chromium blood analysis?

✶ 11. In our first procedure to detect chromium, we converted chromate ions to dichromate. Write down and balance the ionic equation for the production of dichromate, $Cr_2O_7^{2-}$, from reactants—chromate, CrO_4^{2-}, and protons. Water is also a product.

✶ 12. Our next step involves the conversion of dichromate to chromium peroxide, CrO_5 using hydrogen peroxide, H_2O_2. Write down and balance this ionic equation. Two protons are involved as reactants and water is produced.

13. Thioacetamide is once again used as a source of H_2S in our precipitation of Group III metals. Write down this reaction. Assume that the only reactants are thioacetamide, CH_3CSNH_2, and water; and the only products are acetamide CH_3CONH_2, and hydrogen sulfide.

14. Of the four members of Group III that you separate and detect in this chapter, which two have to most acid soluble sulfides?

15. What is the reason for acidifying and boiling the precipitate containing the sulfides of the Group III cations under study?

16. Describe the confirmation test for nickel. What anions are present in this solution? Remember that you dissolved the precipitate in aqua regia.

17. Describe the confirmation test for cobalt. How would you know if you had a negative test?

18. If you saw a deep red color when you attempted the cobalt confirmation test instead of the blue solution expected how would you explain this? What modification could you make to your procedures to allow you to confirm or exclude the presence of cobalt?

19. Describe the confirmation test for iron. How would you know if you had a negative test? Does this test differentiate between Fe(II) and Fe(III)?

20. What is the purpose of the addition of hydrogen peroxide to the mixed iron/chromium solution?

21. Write down the balanced ionic equation for the oxidation of ferrous iron to ferric iron with hydrogen peroxide. What is the other product in this basic solution?

22. What do you think happened to the blue color of the chromium peroxide solution that you created in your first step and set aside in your test tube rack?

23. Why is it appropriate that the final sequences of chromium reactions are carried out as if all chromium present is Cr(VI) and none as Cr(III)?

24. Differentiate between qualitative and quantitative analytical procedures.

25. Differentiate between light absorption and light emission by chemical species. What relative energy states are involved in each?

26. What does the term quantized energy states mean?

27. Why are molecular spectra more complex than atomic spectra?

28. If light in the visible wavelength strikes our eyes what happens?

29. If light outside of the visible wavelength region strikes our eyes what happens?

30. Ultraviolet light is damaging to our eyes and our skin over long periods. Why? Use a comparison between the visible light and UV light in your answer.

31. Why is it so important to pick the wavelength of light used in spectroscopic analysis so carefully?

* 32. What would be the result of setting the spectrometer to a wavelength other that 540 nm in the paint analysis using chromium diphenylcarbazide?

∗ 33. What would happen to the concentration of chromium determined in this test if the calibration curve were prepared at the 540 nm setting and the sample run at an instrument setting of 450 nm?

∗ 34. How are the benefits of double beam spectroscopy taken advantage of in the instrument describe? In other words, how is the double beam effect achieved in the instrument described?

∗ 35. What materials does the reference beam pass through on its way to the detector?

36. How could you construct a single beam instrument and still use a reference to cancel out blank absorption effects?

∗ 37. Spectroscopic cuvettes are usually purchased in matched pairs. Why?

Chapter 5

Barium and Group IV

THE BARIUM MILKSHAKE

The doctor says that it is necessary to get an X-ray image of your digestive tract. It's time for a barium "milk shake." Although it may soon be practical to routinely use an endoscope or nuclear magnetic resonance (NMR) instrument to image this region, it is still common to have an upper gastrointestinal image ("upper GI") generated by using X-ray and an X-ray opaque material like barium sulfate. Your test involves swallowing the barium sulfate suspension, lying down on a table, and having X-ray images generated as the barium moves through your digestive tract. Images of the upper regions including the esophagus and stomach can be simultaneously generated as you lie on the table and drink the suspension through a straw. The barium suspension, which is relatively opaque to the incoming X-rays, provides contrast to the image so that the intestine or duodenum or stomach clearly stands out against the surrounding tissue. In this way doctors can detect damaged or distorted tissue such as intestinal folds or possibly even ulcers.

TOXICITY OF BARIUM

If you are now informed that of the four members of Qual Group IV, barium is the most toxic, what do you think? Would your mind be put at ease if the words <u>very insoluble</u> appeared next to barium sulfate? In fact, barium *is* more toxic than calcium, magnesium, or strontium—the three other members of this group; however, your ingestion of large amounts of a barium sulfate puts you in little danger of the toxic effects of this element. Your knowledge of the aspects of solubility by now provides you with the means to examine this issue with skill and reason: The solubility of barium sulfate is reflected in its K_{sp}, which is 1×10^{-10} and is the reason that the barium sulfate milkshake is a suspension—$BaSO_4$ won't dissolve in water, and it won't move into your body through the walls of your digestive tract either. It is, in effect, merely resident long enough to provide the contrast necessary for the imaging. Finally, as anyone who has had this procedure performed knows, the transit time of this heavy suspension from end to end is often as little as twenty minutes.

The forms of barium that *are* dangerous to humans are therefore the more soluble compounds like barium chloride or nitrate, which are highly toxic. Analogous to barium, magnesium also has a level of toxicity based to a degree on its solubility; yet it too is used in many materials that are routinely ingested, like antacids; like barium, however, injection of *soluble* forms of magnesium can be quite damaging. Magnesium is a metal nutritionally required by animals and is intimately involved in important biochemical processes such as the biosynthesis of proteins, catalysis of ATP (adenine triphosphate) hydrolysis, stabilization of damaged RNA (ribonucleic acid), binding or packing of ribosomal RNA, and interactions with DNA (deoxyribonucleic acid)—all very important processes. The damage that excessive intracellular magnesium can do to a large degree hinges on its effects on these processes and by affecting the flow of ions across the cell membrane. This last process, **osmosis**, is a fundamental aspect of **homeostasis**, the process whereby the body maintains the natural balance of molecules and ions in organs, tissues, and blood. A disruption of homeostasis can lead to toxic build up of elements that are routinely circulated throughout the organism. The toxic effects of both barium and magnesium are exhibited by nausea, as the body tries to regain homeostatic balance and excrete the these elements, respiratory paralysis, and ultimately cardiac arrest as the osmotic balance in nerve cells is disrupted.

CALCIUM AND STRONTIUM

Now that you understand the processes involved in maintaining the order and flow of elements in animals' bodies you should be able to figure out the toxic effects of calcium by yourself. This element is one of the most common elements on our planet (3 % of the earth's crust). It is present in relatively large amount is many different kinds of food and, like magnesium, wide spread in water supplies. Again, like magnesium, calcium is an essential metal and the body uses it in many, many ways. It is essential for bones and teeth, blood clotting, muscle and nerve function, osmotic transport into and out of cells, and maintenance of the cell wall itself. Since this element is so widely dispersed in the body, the route to toxicity must be by intravenous injection or by other means of getting a very high concentration in intimate contact with cells. The result of large amounts of intracellular calcium is osmotic disruption, especially in nerve cells as the flow of other elements like sodium is distorted. This increases the size of the nerve signal necessary to cause the neuron or nerve cell to fire. This causes decreased nerve activity and ultimately stupor. Keeping this in mind, what do you think that an extreme calcium deficit would do?

Strontium presents enough separation difficulties when in the presence of Group IV members that we will avoid it in the analyses below; however, almost everyone who is not color blind is already familiar with this alkali earth metal. Strontium salts are used in the packings of some fireworks. It produces that beautiful rose red color when ignited by an explosion, usually powered by the reaction of a perchlorate oxidizer with sulfur. Just as in inductively coupled plasma, outer electrons in the strontium atoms are promoted to higher energy orbitals by the energy of the explosion, and when they return to ground state they emit a characteristic red light that we all know as the "red star burst."

Strontium's toxicity is similar to calcium but slightly less; where an injected mass of calcium might cause cardiac arrest, the same injected mass of strontium might "only" cause respiratory failure.

Radium and the Curies

In reality, the fifth member of Group IV is the last element in the alkaline earth metals that includes Mg, Ca, Sr, and Ba; however, radium, element number 88, is radioactive, and we will neither separate nor detect it in our Qual scheme. There is an interesting bit of chemical information about radium, however, that does deserve our attention. At the very beginning of this century, experiments involving the recently discovered phenomenon, radioactivity, were going on in many laboratories. After introducing the word radioactivity and in effect founding a new area of chemistry, Marie Curie, her husband Pierre, and their coworkers discovered that when they chemically separated the elements from an ore called pitchblende, they ended up with a "contaminant"—in the barium sulfate fraction—that was radioactive. This element was so chemically similar to barium that their separation procedures—basically large scale dissolution and precipitation steps, like qualitative analysis—did not successfully separate these two elements. With much more careful procedures they did indeed accomplish the complete separation of the compounds in the barium fraction and isolated for the first time 0.1 grams of a radium compound. Though this doesn't sound like much effort they started with two tons of pitchblende and the work took months!

Qualitative Analysis of Group IV Cations

The ordering of the classical Qualitative Analysis Groups is based on which steps are needed to separate, usually by precipitation, the large group of elements that might initially be present in a complex sample. In general, the separations are based on broad classes of precipitation procedures that selectively let soluble compounds pass by (in the supernatant) to be trapped by a subsequent step. The chlorides were the first precipitates (Group I); the sulfides were the second; the hydroxides or slightly soluble sulfides were used to trap Group III; and we will isolate Group IV cations by precipitation as sulfates (SO_4^{2-}), oxalates ($C_2O_4^{2-}$), or phosphates (PO_4^{3-}). Finally there is a group of common ions that will even escape us here: Group V of the next chapter.

Though this book does not use this method, one way to approach this subject is to indeed start with a large group of ions in solution, proceed systematically though the entire Qual scheme, and identify each of the elements that are present along the way. We have instead concentrated on many individual members of each group and simplified the separation and thereby increased the successes at identification.

The three members of Group IV that we will work with, calcium, magnesium, and barium, are relatively soluble as their chlorides, sulfides, and hydroxides and would therefore not be separated very successfully if they were included in any of the previous separations. We will first perform identification tests for all three cations and then proceed to separate each from a complex mixture. The positive test for calcium and barium traditionally involve flame tests—if other interferences are known to be absent. A flame test for strontium is presented because it is so beautiful; however, we will not separate it from the other group members.

Flame tests for barium and strontium

The confirmation for barium after separation will be the precipitation of barium sulfate (insoluble?), and we will augment that by a confirming flame test. If other elements that gave strong flame test were present, like copper or sodium, this test would be inconclusive. We will avoid those interferences.

Put a small amount of barium chloride solid on a watch glass. Break up any clumps so that the $BaCl_2$ is as powdery as possible. Add a few drops of deionized water to the barium compound on the watch glass. Fire polish a Nichrome wire by repeatedly putting it in a Bunsen burner's flame for two or three seconds at a time and alternating this with dips into a 6 M HCl solution.

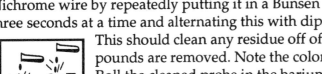

This should clean any residue off of the wire so that interfering compounds are removed. Note the color of the clean probe in the flame. Roll the cleaned probe in the barium solution on the watch glass. Put the end of the wire into the flame of the Bunsen burner. Observe the color of the flame. Barium burns with a yellow-green color. Repeat the test until you have a good idea of the color of barium in the flame.

Be careful not to dump excess $BaCl_2$ solution into the top of the Bunsen burner. Not only is this poor lab technique, it will contaminate your burner and jeopardize your later flame test results.

Repeat your probe cleaning and flame test procedure with a salt of strontium. You may very well not only appreciate the bright color—traditionally referred to as carmine—but you may recognize the flame's color from fireworks displays that you have seen.

Confirmation for calcium

We will separate calcium as an insoluble oxalate and confirm our results by redissolving the precipitate and performing a flame test to get the characteristic orange-red color.

Put ten drops of a 0.5 M solution of calcium chloride in a test tube. Add one drop of 1 M ammonium hydroxide solution and then 1 drop of 0.1 M ammonium oxalate solution. The white precipitate that forms is calcium oxalate. Add more $(NH_4)_2C_2O_4$ solution dropwise until no more CaC_2O_4 comes out of solution. Centrifuge the test tube, decant the supernatant to waste, collect the precipitate, and wash this solid three time with a volume of distilled water equal to the volume of the solid. Remember to stir for a few seconds after adding each portion of water before you centrifuge and decant the wash to the appropriate waste container.

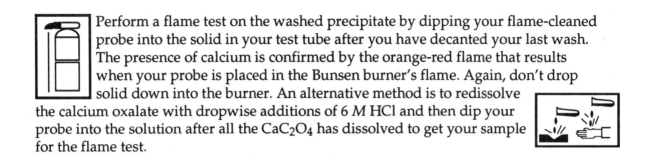

Perform a flame test on the washed precipitate by dipping your flame-cleaned probe into the solid in your test tube after you have decanted your last wash. The presence of calcium is confirmed by the orange-red flame that results when your probe is placed in the Bunsen burner's flame. Again, don't drop solid down into the burner. An alternative method is to redissolve the calcium oxalate with dropwise additions of 6 M HCl and then dip your probe into the solution after all the CaC_2O_4 has dissolved to get your sample for the flame test.

Positive test for magnesium

The positive test for magnesium is based upon the precipitation of magnesium ammonium phosphate and then, if the reagents are available, the subsequent formation of a colored complex of magnesium hydroxide and a magnesium reagent called magneson or S. and O. reagent (after the scientists that developed it). This organic dye turns a precipitate of $Mg(OH)_2$ light blue in basic solution. We will practice the precipitation step with a pure magnesium chloride solution first.

Put 10 drops of 0.5 M magnesium chloride in a test tube. Add 3 drops of 6 M ammonia solution and then 3 drops of saturated sodium phosphate monobasic (Na_2HPO_4). A precipitate of magnesium ammonium phosphate should form in a few minutes. If precipitation is slow or missing altogether, put the test tube in a beaker of boiling water until the tube contents boil. Remove the test tube from the beaker and put it in a test tube rack while you wait for it to precipitate. Repeat the boiling step if no precipitate forms after three minutes. Magnesium ammonium phosphate is a white gelatinous precipitate.

To your washed precipitate add 6 M HCl solution dropwise until the precipitate dissolves. Next add 2 mL of distilled water. Finally add <u>one</u> drop of S. and O. reagent and then slowly add 6 M sodium hydroxide solution drop by drop until the solution is basic to litmus. Add two more drops of base and then observe. The blue "lake" that forms is characteristic of magnesium hydroxide's interaction with the S. and O. reagent and confirms Mg.

Separation and detection of Ba, Ca, and Mg

Get a solution containing a mixture of all three group members (as e.g. the chlorides, approximately 0.1 molar in each). Put ten drop of this solution in a test tube and precipitate the barium as barium sulfate by adding 0.1 M ammonium sulfate one drop at a time until precipitation ceases. You may have to centrifuge briefly between every few drops if the solid does not settle fast enough. After precipitation is complete, decant the supernatant into a test tube marked "Ca + Mg ?" and wash the $BaSO_4$ solid with three 5 mL portions of water.

Confirm barium sulfate as your precipitate by performing the flame test on the solid or by dissolving some of the solid in 1 mL of concentrated hydrochloric acid and dipping your flame probe in that solution.

Take your test tube marked "Ca + Mg ?" and precipitate the calcium that is present by adding ammonium oxalate as you did before. Remember to make the solution basic with a drop of ammonium hydroxide solution *before* you start adding oxalate. After calcium oxalate precipitation is complete decant the supernatant into a test tube marked "Mg ?" and wash the CaC_2O_4 precipitate as before. Confirm the presence of calcium with the flame test. Try the flame test with both your solid and the solid dissolved using sparing amounts of 6 M HCl.

The last test tube may contain magnesium and can be treated just as you did for the $MgCl_2$ standard solution. Perform both the $MgNH_4PO_4$ and $Mg(OH)_2$ confirmations beginning, as before, with 3 drops of 6 M ammonia solution followed by 3 drops of sodium phosphate monobasic (Na_2HPO_4). Your skill at inducing the precipitation and blue lake formation should carry you through to success even though small residual amount of the other Group IV members may still be present. Don't add more than one or two drops of the S. and O. reagent or this solution may hide the blue lake that forms.

Dispose of the material left over after your tests in the appropriate containers. Ask your lab instructor if you do not know where or what the correct container is.

Group IV unknowns

Get a test tube containing a Group IV unknown from your instructor. Record the unknown number if this is applicable. Divide the solution into at least two portions so that you can repeat your separation scheme from the beginning if it becomes necessary. Assume that your unknown contains all the members of this group that we have studied. Write down the steps of the scheme as you perform them making careful notes about the success of each confirmation step. Repeat the known confirmation tests if your are allowed to and if the need arises because of an indistinct result. Report to your instructor your separation scheme and the Group IV members present in your unknown.

Ion Exchange Chromatography

In the very early part of this century, a scientist working in Russia publish a paper that described a means of easily separating the extract of plants into separate components. After pulverizing dried, green leaves and then soaking them in a solvent, Mikhail Tsvet poured a small amount of the solvent mixture into a glass tube (or column) packed with an adsorbent solid (like chalk). When he flowed pure solvent through this packed column under pressure, the plant extract separated into distinct colored bands as it moved down the column (see Figure 5.1). A very powerful, new separation method was born. Since the technique produced individual colored bands of pigment, Tsvet called the process **chromatography** or the chromatographic method (khroma and graphein are Greek for color and to write, respectively). He went on to use 100 different solid column packings and different solvents to separate into individual components a number of different mixtures including chlorophylls and egg yoke. Though other investigators had previously described similar experiments, Tsvet is credited by many as being the founder of modern chromatography.

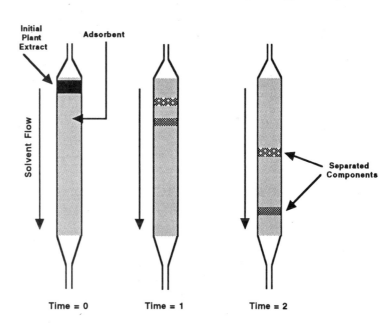

Figure 5.1 The separation of two components by chromatography.

Almost 100 years after the birth of chromatography there are many different kinds of chromatographic separation techniques. Liquid chromatography, gel chromatography, gas chromatography, and paper chromatography are only a few of many different ways to separate components from a mixture—what we have been doing for five chapters. In the second part of this chapter on the Group IV metal cations you will be introduced to another chromatographic method called ion exchange chromatography. If performed correctly, ion exchange allows for the complete separation of a family of ions such as those we have just isolated using the classic Qual scheme. After each fraction is collected, individual elements can be detected using the flame, precipitation, or dyeing procedures that we have just learned. Most often, however, the process of separation is followed by an electronic detector that measures the conductivity (ability to conduct an electric current), refractive index (ability of a fluid to bend light), or UV absorption of the compounds as they exit from the column.

The column

The heart of ion chromatography is the chromatographic column. Most often this is a glass tube filled (packed) with small solid beads called an ion exchange resin of the resin bed. The tube's diameter can be small (< 1 cm) or large depending on the application. The resin, usually made of a synthetic polymer, has a specific chemical structure, called a **functional group** or exchanger, attached to the polymer's surface (see Figure 5.2). The attached functional group is chosen so that it will interact with the ions to be separated, in our case Ba^{2+}, Ca^{2+}, Mg^{2+}, and Sr^{2+}. In Figure 5.2, the exchanger or functional groups are represented as an A^- and are attached to the bead's polymer matrix on the right. Since the functional groups are charged, there is always a **counter-ion** associated with each functional group. For a cation exchange resin (as opposed to an anion exchanger) the counter-ion associated with the resin is often H^+, and therefore the counter-ion in Figure 5.2 is represented as H^+. Before the sample is introduced onto the top of the column, all of the counter-ion sites are filled with H^+ by passing a relatively strong acid solution, like HCl, through the resin bed. This is called regenerating the resin and it assures that all of the functional groups are matched with H^+.

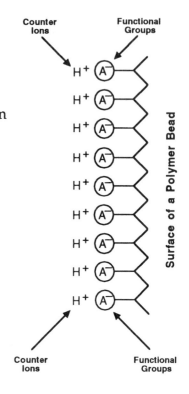

Figure 5.2 Structure of functional groups and ions on an ion exchange resin.

The mobile phase

The liquid that flows through the packing or resin bed from the top of the column to the bottom is called the mobile phase. (Unlike Tsvet's first columns, most ion exchange columns are gravity flow systems—no external pressure is used.) The mobile phase's flow causes the sample separation to occur and then moves the separated components off the end of the column one by one. The major component in the mobile phase is most often water but it also contains a high concentration of the ion exchanger's counter-ion. If this is H^+, as in our example, then the mobile phase can be a diluted acid like HCl.

The mobile phase in introduced onto the top of the column from a reservoir most often situated above the column so that gravity will pull the mobile phase out of the reservoir and through the column. The rate of mobile phase flow is important and must be carefully regulated. One easy way to achieve this is to adjust the position of the mobile phase reservoir relative to the top of the column. Water's desire to "seek its own level" does the rest. If the reservoir is large enough, the flow rate will remain steady for a long enough time to perform the separation.

Loading the column

The process of introducing the sample onto the column is called loading. In our example, a small volume of an aqueous sample of Ba^{2+}, Ca^{2+}, Mg^{2+}, and Sr^{2+} and their counter anions (Cl^- for instance) is added to the top of the chromatographic column by pouring a small amount of the sample solution into the top of the resin bed. These cations (now called co-ions) are attracted to the functional groups of the polymer resin, and, to a large degree, they take the place of the counter-ions on the resin. Since our example cations are doubly charged cations, each co-ion displaces two counter-ions (H^+ ions). Figure 5.3 shows the relationship between the co-ions and the functional groups after the column has been loaded. Each co-ion is related to two functional group positions, and two H^+ ions have been displaced for each co-ion that is adsorbed. The displaced counter-ions are washed off the column with the mobile phase during the next step, called elution or development. The Cl^- anions are not attracted to the like charged functional groups and are therefore also washed out of the column by the mobile phase.

Column elution

The process of separating the loaded cations by flowing the mobile phase through the column is called eluting, column elution, or simply developing the chromatogram, and it exploits an interesting phenomenon: The positive ions that are adsorbed (or bound) to the surface of the polymer are held by their electrical attraction for the negative charge of the functional groups. But each of the co-ions are *adsorbed* onto the resin to slightly different degrees (amounts). Barium cations are bound more tightly than Sr^{2+}, and strontium more strongly than calcium, etc. The result of this differential binding is the key to ion exchange and indeed most forms of chromatography.

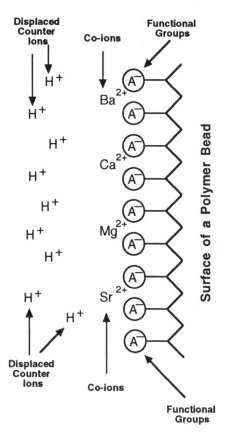

Figure 5.3 Loading the ion exchange column.

When the mobile phase containing a high concentration of H^+ is flowed through the resin bed, the H^+ counter-ions start to reverse the process of loading. They replace the co-ions on the resin; however, the rate at which the counter-ions displace the co-ions depends on how strongly the co-ions are bound to the resin. For the resin and ions that we are discussing, the order of strongest binding to weakest binding is Ba^{2+}, Sr^{2+}, Ca^{2+}, and Mg^{2+}. You might also notice that this order is from the largest to smallest ionic size.

The extent of the co-ions binding depends on a number of factors. A few of these factors are the sizes of the co-ions themselves, their ionic charges, and the kind of functional group on the polymer—many different resins and functional groups are available so this elution order can be different for different resins.

We now have all the components necessary to understand the column elution process: As a mobile phase containing a high concentration of the H^+ counter-ions flows through the resins bed it displaces the co-ions loaded on the bed; however, the co-ions do not elute from the column all at the same time. They come off the column (that is, they are eluted) one at a time in an order that is based on their differential binding relationship with the functions groups on the resin. The result is separation of the individual cationic components of the original mixture. Voilá! Chromatography.

The chromatogram

The flow of mobile phase that exits the column no longer contains the *mixture* of Group IV cations that we loaded on the column. Instead the flow contains pulses or peaks of individual co-ions with periods (in between peaks) of pure mobile phase containing no co-ions. If we carefully collect the column's mobile phase output milliliter by milliliter you can detect the presence of the Group IV cations by the flame tests or precipitation reactions that we studied at the beginning of this chapter. If you collect and quantitatively determine the concentration of the separated cations in each fraction that elutes from the column over time, you are manually generating the **chromatogram**. The word chromatogram originally referred to the separation in space of the compounds on the column itself as it was being developed or eluted; however, in modern chromatographic terms the chromatogram is the recorded result of the chromatographic process: a plot of compound concentration or mass in the exiting mobile phase versus time. In Figure 5.4 the x-axis is calibrated in fraction number and the y-axis in the mass

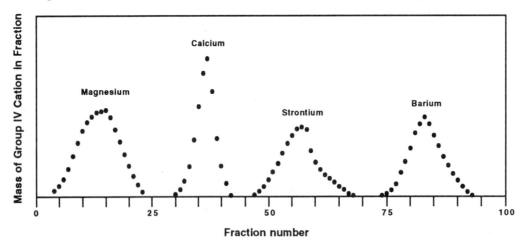

Figure 5.4 Chromatogram of the separation of Qual Group IV cations. X-axis is fraction number. Y-axis is the mass determined in each fraction collected.

of cation detected in that fraction. The fact that the peaks in the chromatogram drop all the way down to the baseline of the x-axis means that each of the individual cations in the original mixture was completely separated from the others; the baseline separation shows that the mass of cations are effectively zero in the fractions at those points.

Another powerful aspect of this kind of chromatography is that the area under the peak reflects the overall amount of that ion in the original mixture; more peak area means that more of that component was present. A smaller peak means less was present. This quantitative information can be used to determine the amount of a particular element in the original mixture.

The conductivity detector

Instead of collecting each individual fraction that elutes from the column by hand in test tubes or in small glass wells, a more powerful and automatic method of generating the chromatogram is to use a conductivity detector. The conductivity detector continuously and electronically determines the change with time of the ionic concentration of the mobile phase as the chromatographic process goes on. This detector sends its signal to an electronic recorder, integrator, or computer. A slight twist is that the H^+ counter-ions present in the mobile phase used to elute the column also conduct electricity. This makes detecting a change in conductivity due to the arrival of the analyte cations difficult (but not impossible—see the experiment at the end of this chapter). A solution is to remove the counter-ions by a small suppressor column placed between the chromatographic column and the conductivity detector. The suppressor column replaces the H^+ with a non-ionic chemical species that does not affect the detector's response. With this device in place, the detector's only response will be to the change in cation (analyte) concentration and the detector's signal will, therefore, only reflect the concentrations of the cations being separated. Figure 5.5 is an exact duplicate of the

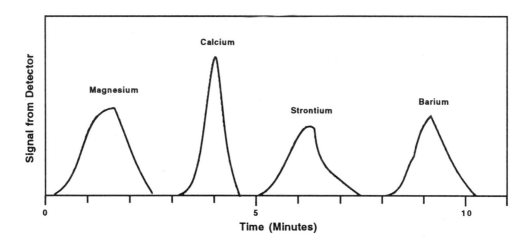

Figure 5.5 Chromatogram of the separation of Qual Group IV cations. X-axis is time. Y-axis is the signal from the conductivity detector.

chromatographic separation performed in Figure 5.4 but this time, instead of plotting individual fractions, collected one milliliter at a time, the x-axis reflects the continuous monitoring of the ionic concentration of the column's output using a conductivity detector. The y-axis reflects conductivity detector signal, and the x-axis is calibrated in units of time (minutes or even hours for some separations).

Standardizing the chromatography

One last point about ion chromatography needs to be made, and this point is true for most kinds of chromatography. The chromatographic separation itself does not tell the analyst what the identity of each peak is as for instance, the formation of the blue lake indicated to you that magnesium was present earlier in the chapter. In most kinds of chromatography, this must be determined by the use of standard compounds. In our example, four separate solutions of each of the four possible ions could be analyzed one at a time, and the four chromatograms generated used to determine the identity of any subsequent unknown peaks. Here's how this process works: The very top of each peak in Figure 5.4 and 5.5 is the highest concentration of that component as it exits the column. For instance, examine the calcium peak that elutes at approximately 4 minutes. Before the peak's top, the concentration of calcium cation that is being detected (in both the sequential fractions and by the conductivity detector) is increasing and after the peak's apex the concentration of calcium is decreasing. The most important aspect of this is that for a particular ion the very top of the peak will always be in the same place on the chromatogram if 1) the column is not overloaded with cations, 2) the mobile phase used to elute is the same and is flowed through the column at approximately the same speed, and 3) the same column is used for the separation. The top of the peak of a compound in a chromatogram is somewhat indicative of the identity of that compound if these conditions are met; therefore, this time receives a special name in chromatography. It is called the **retention time**. For a particular chromatographic separation, retention times are used to identify unknown compounds after known compounds are analyzed to establish their retention times under the same chromatographic conditions used for the analyses of unknowns.

Ion Chromatography Experiment— Separation of Alkaline Earths

The ion chromatographic separation of two alkaline earth metals that we have discussed can be carried out using a cation exchange resin. There are a number of different resins on the market that will accomplish this. Universal Cation resin (Alltech Corp., Houston, TX), PRP-X200 resin (Hamilton Corp., Reno, NV), and IC-Pak C M/D (Waters/Millipore Corp., Milford, MA) are just three of many ion exchange resins that use functional groups designed to separate these types of doubly charged cations. This generalized procedure for the ion exchange separation of magnesium and barium can be applied with or without a counter-ion suppressor column placed after the chromatographic column and with or without a sample injector.

1. The preparation and configuration of the column can be accomplished in many ways. These details will be left to your instructor; however, the column must be packed and pre-rinsed (regenerated) in such a manner that the length is sufficient to separate magnesium from barium and will produce a flow rate that is acceptable for the time allotted for the experiment. Two mL/min is a normal flow rate for a 1.25 cm diameter ion exchange column. Finally the regeneration of the column must be performed in such a manner that the column's function groups are all matched with an appropriate counter-ion, usually H^+. This can be accomplished by making the final rinse a dilute HCl or HNO_3 solution; however, follow the manufacturer's instruction in these steps.

2. If a suppressor column is being used then make sure that column is correctly prepared to suppress the ions of the eluting mobile phase. If a membrane suppressor is being used then carefully follow the manufacturer's procedures for its use.

3. Prepare a mixture of magnesium and barium ions (from their chloride salts) such that the concentration in the mixture is 25 to 50 µg/mL (25 to 50 ppm). Depending on the elution procedure, controlling the acid content of the sample solution may also be necessary. A total volume of 100 mL of solution will allow you enough sample to perform multiple runs.

4. Place the necessary eluent solution (mobile phase) in the solvent reservoir. In its simplest form, this solution may simply be a dilute acid like HCl; however, it may contain a complexation reagent like ethylenediamine tetracetate (EDTA) or phenylenediamine hydrochloride. The choice of these reagent depend on the column resin and detection system that you use.

5. Establish the background (baseline) plot for the detection system following the manufacturer's directions. A flat baseline from the detector's recorder means that the detector has stabilized and that you can make an injection. A rising or falling baseline must normally be stabilized before an injection.

6. If you are using the sample injector, turn the injector's handle to its fill position and flush the sample loop two times with the sample solution using a clean HPLC syringe. Inject the volume of the sample loop onto the column by turning the injector's handle to the injection position. Typical loop volumes are 0.1 to 0.25 mL (100 to 250 µL). The injector automatically, yet temporarily, halts the flow of mobile phase to accomplish the injection and then re-establishes eluent flow as the chromatographic run begins. You may need to mark the chromatogram manually to denote the time of your injection.

7. If you are manually loading the sample, place an appropriate volume of sample solution, usually milliliters, at the head of the column using a pipette. Make sure that the resin bed in never uncovered by solution. Start the flow of the mobile phase and mark the chromatogram to note the time of injection if this is necessary.

8. As the chromatogram develops you will see two individual peaks elute from the column. If you are not using a suppressor column you may see another substantial peak that elutes very early before magnesium and barium. This is caused by the quick elution off the column of a high concentration of counter-ions that were displaced by the barium and magnesium as they were adsorbed by the ion exchange resin. This peak, sometimes confusingly referred to as the "solvent" peak, may be very large but should be very sharp. The term solvent peak is a carry over from other chromatographic methods, specifically gas or liquid chromatography.

   If you are using the suppressor column and the concentration of the analyte cations is not too great then the only peaks that you will see will be first magnesium and then barium. If the chromatography using the suppressor column is unsatisfactory (ask your instructor) then one alternative may be to decrease the concentration of the analyte cations in the sample solution by a factor of 5 or 10 and then to repeat the run after regenerating the resin bed.

9. After you have obtained a successful chromatographic separation you will need to confirm the identity of your peaks. After flushing and regenerating the column as before, inject a solution the contains only magnesium at the same concentration as in your standard mixture and record the chromatogram. Again, after regenerating the column, repeat with a solution only containing barium. The retention time of these standards will coincide with two of the peaks in your mixture's chromatogram.

10. The final step in this procedure may be the regeneration and storage of the ion exchange resin. As stated many time before, follow the manufacturer's direction in this regard. These resins are not inexpensive materials and should only be disposed off at the direction of your instructor.

11. Make sure that the solutions that you have produced in your analyses are disposed of in the appropriate container. *If you are not sure where this container is, ask someone who knows!*

CHAPTER FIVE
PROBLEMS

1. Is the patient in danger of being poisoned by toxic barium when undergoing the procedure that uses the barium milkshake? Explain.

2. What measure of barium sulfate's solubility can you use to support your explanation of the answer to problem 1?

3. Can you estimate the solubility of radium sulfate from the description of how this element was discovered? How?

4. Given that the order of precipitation reactions in all of the Qual scheme so far has been Group I—chlorides, Group II—sulfides, and Group III—hydroxides, order the solubility of $BaSO_4$, BaS, $BaCl_2$, and $Ba(OH)_2$ from most soluble to least soluble.

5. The flame test for barium can be complicated by contamination of elements like copper or sodium. Why?

6. If you heat a clean glass rod in the flame from a Bunsen burner long enough you will see the same color that you get when you perform a flame test with table salt. Why?

7. If you strike most old Bunsen burners against the table top while they are lit you will see a range of bright colors in the flame. Why?

8. The effects detected in problem 7 even occur with a scrupulously clean Bunsen burner if it is old enough. Why?

9. Why is it necessary to use hydrochloric acid to pre-clean the glass rod or Nichrome wire *before* it is used in the flame test? Why not use sulfuric acid?

10. Write down the complete ionic equation for the reaction of sodium sulfate and barium chloride. What drives this reaction to completion?

11. Write down the ionic equation for the reaction of ammonium oxalate with calcium chloride. Why do you think that a drop of ammonium hydroxide is added to the test tube before ammonium oxalate is added?

12. The blue lake formed with magnesium hydroxide uses a reagent traditionally called the S. and O. reagent. Where does this name come from?

13. Draw a flow chart of the separation of each of the components in a mixture of Group IV cations if the original mixture contains only barium.

14. Repeat problem 13 for a mixture of barium and calcium.

15. Repeat problem 13 for a mixture of all three cations that we separated in this section.

✱ 16. After the S. and O. dye confirmation test, how could you recover the contained magnesium *free of the S. and O. reagent*, that is a pure undyed precipitate?

✱ 17. One of the separations that we avoided was separation of Ca^{2+} and Sr^{2+} using 16 M nitric acid. The precipitate formed is $Sr(NO_3)_2$. If collected and washed, this precipitate should contain strontium. How could you prove this and how could you precipitate and confirm calcium in the highly acidic supernatant?

✱ 18. If you started with a mixture of barium, calcium, and magnesium chlorides and initiated the precipitation of $BaSO_4$ by using sulfuric acid instead of ammonium sulfate, how would this affect the subsequent steps in your efforts to separate the remaining cations? Be specific.

✱ 19. Precipitation for "stubborn" solutions can sometimes be initiated by using a stirring rod to scratch the bottom of the inside of the test tube containing the solution or by adding a small crystal of the precipitate (from a laboratory bottle of the dry solid). Explain why each of these laboratory "tricks" work?

20. Where does the analytical process called chromatography get its name?

21. Using a dictionary, differentiate between adsorption and absorption.

22. What is the attraction between two ions with an opposite charge and what is the attraction between the negatively charged functional groups and the counter-ions in an ion exchange chromatographic resin?

23. What is the attraction between the co-ions and the functional groups in an ion exchange chromatographic resin?

24. When a new ion exchange resin of the nature that we have been discussing is received fresh from the manufacturer it is always washed with a strong acid before it is loaded with the analyte to be separated. Why?

✱ 25. Describe how you can regulate gravity flow to maintain a relatively constant flow rate through an ion exchange column without the means of a solvent pump.

✱ 26. The mobile phase that we used to elute the cations from the column was dilute hydrochloric acid. You could not use sulfuric acid for this same chromatographic separation? Why not?

27. Each co-ion that was loaded on our ion exchange column "occupies" two functional groups. Why?

28. What happens to the counter-ions that were displaced by the co-ions?

29. What are the axes in the two chromatograms that you were introduced to in this chapter?

* 30. Instead of fraction number, sometimes a unit called the elution volume is plotted on the x-axis. How could you calculate the elution volume at which the apex of the calcium peak elutes from this column?

31. Since the conductivity detector responds to changes in ionic strength, the large amount of H^+ in the mobile phase must (often) be removed in order for the detector to correctly track the change in ionic strength of the cations that are being separated. How can this be accomplished in ion exchange chromatography?

32. How are the peaks that are plotted on the chromatogram identified in this and most kinds of chromatography?

Chapter 6

Group V
Acid Rain and The Ion Pump

INTRODUCTION

The three cations left in our trip through the classical Qualitative analysis scheme are ammonium, potassium, and sodium. Unlike all the cations we have studied so far (except the Hg_2^{2+} dimer), ammonium is a polyatomic cation, NH_4^+ and, like both K^+ and Na^+, it is found everywhere in our environment. All three of these species play many important roles in the environment and in biological systems. You will be introduced to some of the environmental chemistry of the Group V cations before their qualitative analysis is detailed.

AMMONIUM EXCRETION AND ACID RAIN

The role that ammonium cation plays in our lives starts in our bodies and ultimately influences not only our immediate surroundings but to some degree even the pH of the atmospheric layer of air closest to the earth. Nitrogen atoms are a significant component of all amino acids, the molecular precursors of proteins. Since excess ingested amino acids cannot be stored, the nitrogen atoms in them are either recycled in the biosynthesis of other nitrogen containing molecules or the excess nitrogen is excreted from the body in the waste stream. If this process goes awry it can be very dangerous for the organism. In humans, a condition that causes lethargy and in extreme cases retardation has been documented and labeled **hyperammonemia**. It involves a high concentration of ammonium ion in the blood because of a defective excretion cycle possibly originating in a damaged liver.

Most animals release excess nitrogen as urea or uric acid; however, many aquatic animals just simply discharge NH_4^+ to keep their waste cycles in balance. This means that one way or the other, NH_4^+ is a significant component of most animal wastes. As almost anyone knows who has been near an area where a large collection of animals lives or near a few days old diaper pail, one of the major degradation components released by these compounds in the environment is ammonia, NH_3. In fact, a major source of gaseous ammonia into the atmosphere is from animal wastes; however, there are other sources as well. Figure 6.1 on the next page shows a generalized diagram for the ammonium cycle. The important processes involve ammonium production. This includes the decomposition of organisms (including their wastes), nitrogen fixation and

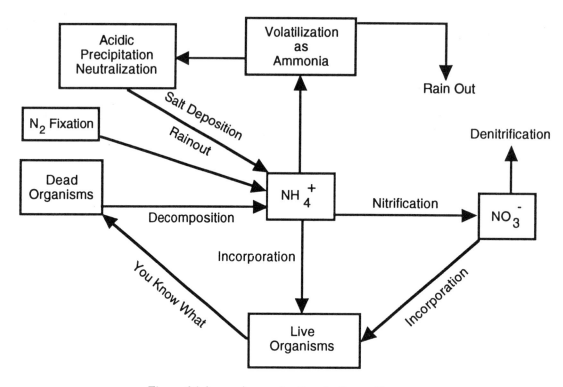

Figure 6.1 Ammonium cycle of production and loss.

nitrification by bacteria, and deposition and rainout of ammonium salts. These sources can be contrasted with the prominent ammonium losses via denitrification and organismic uptake and incorporation.

The role of ammonia once it escapes into the air is unique among atmospheric constituents. Ammonia is the only basic gaseous component in the atmosphere; that is, basic in a pH sense. This is in contrast to the many acidic components present in the air surrounding the earth: Nitric acid, sulfuric acid, hydrochloric acid, and carbonic acid all play significant roles in the pH of clouds and the precipitation that falls to the earth. Low **pH** precipitation (pH <5) is called **acid rain** or acid snow and both of these affect soil, water and wildlife. Some lakes in the Northeastern United States, Eastern Canada, and Scandinavia have pH readings 10 to 50 times more acidic than their normal pH would be without the effects of acidic precipitation. This problem is vastly complicated by the influence that the soil and underlying rock play in their interaction with the deposited acids. Some acid deposition can be partially neutralized by the ground and some can't. This means that some geological areas fair better than others even though the pH of the precipitation is still much lower than normal. Geology plays a large role in the ultimate effects of acid rain.

Gaseous ammonia can react with atmospheric acids and neutralize them in cloud water or on small particles. And as noted above, NH_3 is the only gas that can do this in the atmosphere. The products of this neutralization interaction are rained out as salts like ammonium sulfate, ammonium chloride, or ammonium nitrate. This can, therefore, short circuit the acid rain problem before it touches the ground.

Unfortunately the large amount of gaseous acids in the Northern hemisphere, created to a large degree by the release of sulfur and nitrogen oxides from the combustion of fossil fuels, is *not* completely neutralized by ammonia, and environmental damage varies from region to region. The effects of acid rain are wide spread in many of the industrialized nations and their down-wind neighbors: many lakes can no longer support fish; plant and crop damage has been documented; and forest damage in Germany may even terminally threaten the Black Forest. Experiments have been underway for many years in lakes in Scandinavia, with varying success, designed to neutralize the acid in natural lakes by the addition of neutralization reagents.

POTASSIUM, SODIUM AND YOUR NERVES

Potassium metal, K^0, and sodium metal, Na^0, are so intent on losing their lone s electrons and thereby so reactive that they will react with water to produce elemental hydrogen. This reaction can be used as a spectacular demonstration by tossing a small piece of either metal into a pan of water. Because the hydrogen gas produced often catches on fire, the reacting metal runs around on top of the water continuously reacting and putting out a flame to boot! However, the reactive characteristic of potassium and sodium are actually an unusual episode in the life of an average potassium or sodium metal. Reactivity of these pure metals is such that very little of these species exists naturally. Instead the most common state of these elements is missing that last valence electron and forming the +1 ion. Furthermore, the solubility of compounds containing potassium and sodium are so great that these ion are found almost everywhere in the environment. Like sodium, the distribution of potassium on the earth is approximately 2 to 3 %. And finally, the great solubility of these compounds means that we will not use a precipitation reaction as a means to isolate potassium or sodium from a mixture of our Group V cations. Instead we will depend, once again, on flame tests. The complication of overlapping light emissions from these species will be discussed below.

Potassium and sodium ions not only take the role of escorting dissolved anions around in our environment, they also play an important part in the ionic pumps in our bodies. Their job of controlling the ionic potential across neural membranes is extremely important and was eluded to in our discussion of calcium imbalance and homeostasis in the last chapter.

The sensation of touch felt by your finger is actually an electronic signal transmitted via nerve cells or axons that are bundle together by tissue into a structure we call a nerve. The electronic signals flowing through nerve cells, passing information to the central nervous system, are actually transmitted by shifting the distribution of ions between the interior and exterior of the axon itself. The relative distribution of sodium and potassium on either side of the cell wall directly affects the amount of signal that will cause the nerve cell to "fire."

Through a series of biochemical reactions, cell walls have the ability to move or pump ions (and other species) from one side of the wall to another. In the resting or unexcited state, the nerve cell's success at pumping K^+ and Na^+ is such that sodium preferentially builds up outside the cell and potassium inside. Since each of these cations is normally balanced electrostatically by a counter-ion (think ion exchange chromatography), this preferential buildup means that the inside of the axon has an excess negative charge relative to the outside (usually about -70 millivolts). This electrochemical energy gradient or potential means that the cell is, in effect, cocked and ready to fire. This electrochemical potential is called the resting potential.

If the axon's potential is disturbed strongly enough the result is a chain reaction that triggers the electrical disruption of all the axons along the nerve bundle, that is, serial firing. We call this traveling electrochemical event a nerve signal or nerve impulse, and the stimulus for it can be a touch on the skin or a signal from a nearby nerve.

Now you may see the importance of maintaining the correct concentration of cations in and around cells. If the wrong concentration of a cation like calcium distorts the electrochemical gradient of the nerve cell, the result is a change in the resting potential. A *lower* resting potential means that it takes less of a disruption on the nearby axon to cause the cell to fire. The result is that the nerve either fires too easily or even fires with no stimulus at all. This could lead to a very dangerous situation if the job of that nerve is to tell the heart muscle when to contract.

If the resting potential is *increased* by the chemical imbalance then the result is an inhibition of nerve impulses and lethargy, stupor, or possibly even death as the lungs stop working or the heart stops pumping blood.

Atomic Emissions

Remember that we mentioned above that the flame tests for sodium and potassium were a little more complicated than usual? In our past tests we simply excluded interfering compounds that were not in the group under study so that we could clearly see the emission features (think colors) of our cations: strontium—carmine, calcium—orange-red, barium—yellow-green, and tin—blue. With the Group V members this is impossible because both potassium and sodium are in the same group. Since we are not going to use precipitation we must be able to distinguish between the light given off by sodium and potassium excited in a flame. Figure 6.2 has two parts: 6.2a and 6.2b are the *emission* spectra in the visible region of sodium and potassium in a flame.

The light given off by sodium has basically one strong emission line in the visible, at about 589.3 nm (actually two closely spaced emission lines above and below this point). Potassium has two less intense lines at 405 nm and approximately 768.5 nm (this last wavelength is again two closely spaced lines). The result of a sodium flame test is a bright yellow light.

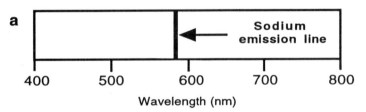

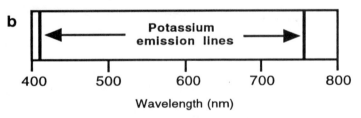

Figure 6.2 Emission spectra of sodium and potassium.

The result of a potassium flame test is a weaker reddish-violet light that unfortunately is completely masked by the emission of almost any sodium if it is present. What can you do to resolve this dilemma? If you said a monochromator you would in fact be correct. If you had an atomic absorption spectrometer you could indeed use that instrument (and its flame) to determine the presence of either of these species using its monochromator to isolate the elemental emission lines that you desired without having to use a hollow cathode lamp. (This would actually be atomic emission spectroscopy.) But there is an easier method of qualitatively solving this problem that has been used for years: a simple filter.

Figure 6.3 shows the wavelength region that is transmitted by cobalt blue glass. Remember, this is the transmission spectrum, that is, a plot of light that will be allowed to pass through the filter. Notice that "shoulders" at either end of the filter's transmission spectrum allow potassium's emission lines to pass to a large degree and that the flat center portion of lowest transmission blocks the sodium emission line. If sodium and potassium are present in the same sample, examining the flame without the filter will undoubtedly reveal the presence of Na since its emission line is so intense. When it comes time to test for potassium, with the filter between your eyes and the flame almost all of this sodium emission will be blocked and you will see some light that will indicate the presence of potassium. Though this one takes a little practice, working with the filter can be fun, and it is substantially cheaper than a $20,000 spectrometer!

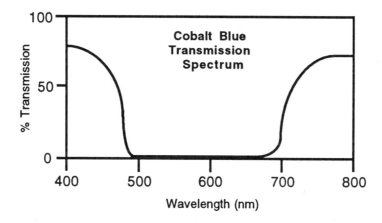

Figure 6.3 Transmission spectrum of cobalt blue glass.

Qualitative Analysis of Group V Cations

We are, in general, not approaching the entire Qualitative Analysis scheme by starting with a mixture of all of the possible dissolved cations and then separating each out for identification; however, even this book's abridged scheme involves fourteen or fifteen elements so far, so you could imagine how challenging that task might be in the classical scheme. This chapter focuses on those cations that would traditionally slip through all of the other Qual Group analyses of selective dissolution, oxidation, decantation, reconcentration, and precipitation and that finally end up in the last Qual Group, Group V, the soluble cations. Since environmental samples almost invariably contain potassium and sodium (see above) these analyses have, in general, only passing importance in environmental chemistry. The importance of ammonium determination is more significant in environmental samples such as those collected from precipitation; however, that analysis undoubtedly involves a quantitative measure not a qualitative one because of the ubiquity of this ion in the atmosphere.

The potassium and sodium analyses that you will use are, as detailed above, flame tests, augmented by the use of a glass filter to isolate the potassium emission. The ammonium test is likewise relatively simple. It involves the only odor test in this book and is backed up by the detection of very small amounts of ammonium by its conversion to ammonia with strong base and its absorption and subsequent color change on litmus paper.

Sodium and potassium flame tests

You will need a Nichrome wire flame probe for these tests; a glass rod won't work as the flame probe. Why?

Dissolve a small amount of sodium chloride in 1 mL of distilled water. Dip your pre-cleaned Nichrome wire into the solution and place it in the flame. The bright yellow color of sodium will be prominent. Repeat this test three or four times. On one run view the emission through one thickness of cobalt blue glass. Next view the sodium emission through two thicknesses of cobalt blue glass. The emission level will be substantially attenuated (decreased) by the filter.

Dissolve a small amount of potassium chloride in 5 mL of distilled water. Clean your flame probe with repeated heatings and dipping in distilled water until no more sodium emissions are detectable. Dip your pre-cleaned Nichrome wire into the KCl solution and place it in the flame. The reddish-violet flame of potassium will be evident, especially if you put the probe in the lower, less blue part of the methane cone. Potassium's emission does not last as long as sodium's so you will have to repeat this test many time to become familiar with its color. After this, repeat the test a few more times using the cobalt blue glass. The attenuation that you noticed with the sodium line should be substantially less with this test because of the transmission characteristics of the cobalt blue filter (see Figure 6.3).

Try different mixtures of solutions of the potassium and sodium salts and perform the flame test as described above. Start with a pure potassium solution and beginning with a single drop, add larger and larger volumes of the sodium chloride solution to the test tube containing the potassium solution. Perform the flame tests between each addition. Your addition of a sodium salt solution to the potassium solution achieves two things. It adds sodium, and it dilutes the potassium concentration. You will probably find that a large enough concentration of sodium cation can successful swamp potassium's emission line even with the filter in place.

Ammonium/ammonia odor test

If this is your first smell test follow the instructions exactly as they are written. *You are responsible for your own safety in the laboratory*, and while a strong whiff of ammonia is not permanently damaging, it is quite unpleasant.

Put 2 drops of stock 1 M NH_4Cl solution in a test tube. Add a drop of 6 M NaOH. Do not let this aqueous base touch the upper part of the test tube. Make sure the drop falls all the way to the bottom of the test tube. Carefully—very carefully—hold the test tube in front of you over the lab bench twelve inches or more from your nose. Slowly fan your hand over the test tube and toward your nose. Continue this process, slowly moving and fanning the test tube toward your nose until you smell ammonia. Put the test tube in your test tube rack.

Tear a one inch piece of red litmus paper in half and moisten one piece with distilled water. Put the moist litmus paper inside the top of the test tube and stick it against the glass. If the ammonia being released is in high enough concentration the litmus will turn blue. Red litmus turns blue in contact with base.

Dilute your stock 1 M NH_4Cl solution by a factor of ten with distilled water by adding two drops of the stock solution to 18 drops of distilled water in a clean test tube. Label this solution "0.1 M NH_4^+." Repeat the odor and litmus test in a clean test tube with two drops of this newly diluted solution.

Continue diluting each solution that you test by a factor of 1:2 into clean test tubes. A typical 1:2 dilution is four drops of sample to four drops of distilled water; but the portions can be varied. Don't forget to stir each tube with a few finger taps near the bottom.

Perform the odor test until you can no longer smell the ammonia given off. Clearly label each tube with its new concentration. For very dilute solutions to get a positive litmus test you may need to heat your test tube in a Bunsen burner's flame for a few seconds after the NaOH addition to volatilize enough ammonia to cause the litmus to react. What is the lowest concentration that you can smell? What is the lowest concentration that will cause a positive litmus test?

Dispose of the material left over after your tests in the appropriate containers. Ask your lab instructor if you do not know where or what the correct container is. Also make sure that your Bunsen burner is free of any dried salts that you might have spilled accidently.

Group V unknowns

Get a test tube containing a Group V unknown from your instructor. Record the unknown number if this is applicable. Assume that your unknown contains all the members of this group. Perform the flame and odor test on small portions of your unknown solution. Repeat the known confirmation tests if your are allowed to and if the need arises because of indistinct results from the flame tests. Report to your instructor the Group V members present in your unknown.

Group V Detection with Ion Selective Electrodes

Hydrogen ion does not fall into any of the Qual Groups that we have studied. Even the "soluble cations" of Group V do not include hydrogen. This is because of the role that hydrogen cations play in aqueous chemistry. This species is so wide spread and important in aqueous systems that its concentration is routinely measured all the time and most often reported in a somewhat special way, as the negative log of the hydrogen ion concentration, pH. The measurement of hydrogen ion concentration, denoted as [H^+], is so important that specials instruments are used to determine the pH of a solution. Furthermore the regularity and importance of the measurement of this ion also means that its determination can not be a difficult process. And indeed it isn't. While litmus paper can help you report this value in a general way, a pH electrode is used for more exact determination of pH.

The pH electrode

Figure 6.4 shows a drawing of a pH and reference electrode that respond to hydrogen ion concentration in a solution in which the electrodes are placed. There are a few things to note. The measuring or indicating electrode on the left includes a thin glass membrane at its tip that will allow H^+ to pass through. Inside this electrode is a silver wire coated with silver chloride and dipping into an internal solution of dilute HCl. The reference electrode on the right is constructed so that the electromotive force (EMF) or cell potential generated by the relationship between the measuring and reference electrodes electronically translates directly into a standard value that can be used to determine [H^+], hydrogen ion concentration.

The difference between [H^+] in the measuring electrode and outside the membrane is like the resting potential of the axon discussed earlier in this chapter. A gradient of ion concentration across the porous membrane creates an electrical potential. If the potential is due to the concentration of only one ion and the internal solution's ion concentration is fixed, measuring this electrical potential is actually a measure of the ion's concentration in the bulk solution. The electrical potential across the porous membrane is electronically displayed by a voltmeter that is fed signals from both electrodes, measuring and reference. Instead of displaying millivolts of electrical potential, the pH meter can be programmed to report directly in pH. The pH meter is standard equipment in most laboratories.

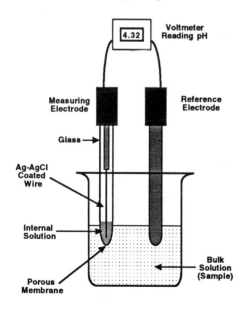

Figure 6.4 Measuring pH and reference electrodes.

The birth of ion selectivity

Now remember what happens when the ion concentration gradient is distorted by an interfering ion; in our nerve's axon example it was by high concentrations of Ca^{2+}. The result was that the cell's resting potential changed and this changed the response characteristics of the nerve cell. In the early work with glass pH electrodes it was discovered that the composition of the glass membrane had a large effect on the pH measurement. Something in the glass was interfering with the potential developed across the membrane. This research led to new membranes that instead of responding to [H^+] and reporting pH, respond to the concentration of other ions—they selectively detect one ion in the presence of another. The ion selective electrode was born!

The solid state ion selective electrode

One design of an ion selective electrode is constructed quite similarly to a pH measuring electrode except that the membrane is made of a material, sometimes a polycrystalline solid, that permits the establishment of a potential gradient between the internal solution and the ionic concentration of a particular ion outside in the bulk solution. In other words it conducts specific ions. Figure 6.5 displays an example of this design. It shows the indicating electrode only; a reference electrode similar to that in Figure 6.4 is also connected to the voltmeter but is not shown.

The makeup of the internal solution in this indicating electrode depends on the ion to be detected in the bulk solution. Within a relatively wide concentration range of, for instance, potassium ion, a potassium indicating electrode's electrical potential linearly reflects the concentration of that ion in a solution in contact with the membrane's surface. Just as before, if the concentration of interfering ions is too large the readings can be distorted.

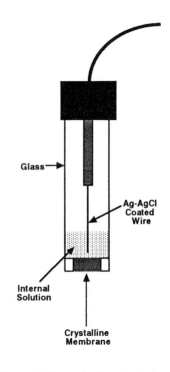

Figure 6.5 Ion selective electrode.

A liquid ion exchange membrane electrode

Another design of an ion selective electrode involves an ion exchange solution somewhat similar to the resin used in Chapter Five. The ion exchanger modifies the porous membrane making it ion specific. Figure 6.6 on the next page is a drawing of a liquid ion exchange measuring electrode. This design is only one more step more complicated than the solid state electrode in Figure 6.5. The edges of the porous membrane are in contact with an ion exchanger made of an organic (water insoluble) solution. The makeup of the exchanger is chosen carefully so that the transport of a specific ion, for instance, ammonium, across the membrane is facilitated by the ion exchanger and, simultaneously, interfering ion transport is hindered. The more successfully that this is accomplished, the more selectively the electrode will respond and the higher the concentration of interfering species can be without distorting the readings.

Applications of ion selective electrodes

Ion specific electrodes to measure cations like Ag^+, Ca^+, Cs^+, K^+, Li^+, Mg^+, Na^+, NH_4^+, Rb^+ have at one time been available commercially. Anionic specific electrodes have also been developed to determine bromide, carbonate, chloride, cyanide, fluoride, iodide, nitrate, and sulfide.

These specialized electrodes have a wide range of applicability in the medical and environmental fields—measurement of ammonium ions in acid rain or snow being only one of many worth mentioning specifically. Another is the *in situ* measurement of metabolic constituents in an effort to probe the function of living systems. In 1991 a Nobel Prize was awarded to two scientists who used ion specific microelectrodes to probe the iron ion pump in living cell membranes—the general area that we discussed at the beginning of this chapter in the area of potassium and sodium concentration gradients.

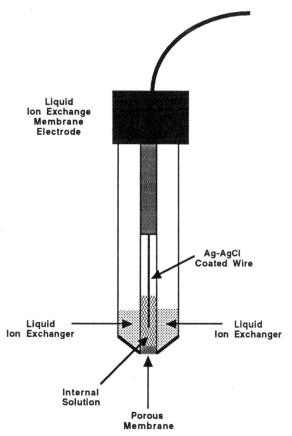

Figure 6.6 Liquid ion exchange membrane electrode.

ION SELECTIVE ELECTRODE EXPERIMENT— FLUORIDE DETERMINATION

The determination of fluoride anion in treated public water systems is a quick and easy way to demonstrate the use of an ion selective electrode. After producing a calibration curve for a range of known fluoride ion concentrations, the concentration of fluoride in an unknown or a tap water sample can be determined in a manner similar to the dichromate experiment described at the end of Chapter 4. The requisite fluoride selective electrodes are sold by a number of manufacturers: Fisher Scientific and Orion are two widely accessible vendors. Using a buffer solution to control the total amount of ions in each solution that is examined, this system as described will respond, therefore, only to the concentration of F- in the standards and in the unknown solution. The calibration curve that is generated by the standards is plotted using semilog paper (log x-axis). And as in Chapter 4, the concentration of the fluoride ion concentration is determined from the calibration curve.

The buffer required for this experiment is called total ion strength adjustment buffer (TISAB) and is commercially available from Fisher Scientific (Houston, TX) or Orion Research (Boston, MA). This compound is added to all of the standards and samples.

1. Make sure that the fluoride selective electrode has been soaking in a solution containing 100 ppm fluoride ion for at least one hour before the analysis. Also soak the reference electrode in a pH 7.0 buffer solution for an hour before analysis. Unacceptable drift when taking the readings proves that the electrodes has not been prepared correctly.

2. Using a standard solution of fluoride ion (usually 100 ppm NaF), make three standards containing 1.0, 10.0, and 25.0 ppm fluoride ion in solution: Pipetting the appropriate amount of the standard into a 100 mL volumetric; add 50 mL of TISAB buffer to each, and make each of the three standard solutions up to the same final volume, 100 mL, with deionized water. As long as TISAB is a substantial component of each solution, its ionic concentration will dominant the ionic strength of all solutions, and all will have a pH between 5 and 6.

3. Prepare the fluoride containing water sample in the same manner as the standards. Make sure that TISAB is present in the same approximate volume. Keep track of your dilution of the water sample in order to back calculate the fluoride content in that solution from the calibration curve correctly.

4. Use the fluoride selective electrode and the reference electrode (usually saturated calomel) to measure the electrical potential of each of the standards and the sample: With each solution in a separate beaker, place these electrodes in each standard starting with the least concentrated. Stir the solution for two or three minutes and record the potential for each.

5. Some meters allow the analyst to enter the concentration of each standard measured and then subsequently displays the result of an unknown measurement directly as fluoride concentration. If this is possible, follow the manufacturer's calibration procedure and report the concentration of fluoride in the unknown, *taking into account the dilution of the sample that you performed above.*

6. If the system does not allow direct concentration measurement, plot the standard's potentials (in millivolts) on a piece of graph paper with a linear y-axis and the associated fluoride concentrations on a log x-axis. Report the concentration of fluoride in the unknown, *taking into account the dilution of the sample that you performed above.*

7. Make sure that the solutions that you have produced in your analyses are disposed of in the appropriate container. *If you are not sure where this container is, ask someone who knows!*

CHAPTER SIX
PROBLEMS

1. What is the source of the nitrogen that is excreted by animals?

2. What is hyperammonemia?

3. What is the only basic gaseous component of the atmosphere? What is its formula?

4. Write down the names and formulae of three acidic atmospheric components.

5. Write down the balanced neutralization reaction between sulfuric acid and ammonium hydroxide.

6. Why does a low pH rain not always result in a low pH runoff?

∗ 7. Most of the lowest pH rain that has been detected is in the Northern hemisphere. Why?

∗ 8. Assuming that the wind flow in the Northern hemisphere is from west to east, use a map to determine the source of most of Scandinavia's acid rain assuming that the power plants that release SO_2 and oxides of nitrogen are not in Scandinavia. Can you do the same for New York State?

9. Why can't you routinely use a precipitation reaction to isolate sodium or potassium as we could with so many of the earlier group members?

10. Write down the definition for an axon's resting potential.

11. Which disruption in resting potential causes an increased "case of the nerves."

12. What is the wavelength of sodium's single emission line? What color is this light?

13. What are the wavelengths of potassium two emission lines? What colors are these? Why do you think that the color of the potassium flame is purplish?

14. What wavelength regions does a cobalt blue filter transmit best?

15. What does the transmission spectrum of window glass look like?

∗ 16. What does the *absorption* spectrum of a cobalt blue filter look like?

17. The sodium flame test cannot be performed using a glass rod as a flame probe. Why?

18. The potassium flame test cannot be performed using a glass rod as a flame probe. Why?

19. Why is the potassium emission so easily swamped by such a low concentration of sodium in your mixed Na/K salt test?

20. Write down the reaction between sodium hydroxide and ammonium chloride that produces ammonia. You may write this as a two step reaction if you choose.

21. Would another base perform the same job that sodium hydroxide does in the ammonium detection scheme?

* 22. Why do you think that it is necessary to heat a sample that contains a very low concentration of ammonium to get a positive test? Use the equation(s) in your last answer as support.

23. What is the driving force in the reaction between ammonium chloride and sodium hydroxide that takes this reaction toward completion: precipitation, formation of water, or formation of a gas?

24. What is the definition of pH? Write down this definition and calculate the hydrogen ion concentration in the solution in the beaker in Figure 6.4.

25. What is the purpose of the permeable (porous) membrane in a normal pH electrode?

26. What is the purpose of the internal solution in a normal pH electrode?

* 27. Why do most pH meters also allow you to switch them to an output in millivolts?

28. What kind of error led to the discovery of ion selective membranes?

29. Why is it important that the response of an ion selective electrode's potential be linear?

* 30. If an ion selective electrode's response in not linear (see problem 29) but the response curve is known, can the instrument still be useful as an ion selective electrode? How?

31. What is the purpose of the ion exchange liquid in the liquid ion exchange membrane electrode?

* 32. A urea selective electrode has been developed that uses a gel coating containing an organic chemical that hydrolyzes urea to ammonia. NH_3 then forms NH_4^+ in contact with water in the gel. Therefore NH_4^+ is the actual chemical species detected by the electrode. Make a drawing of how this urea detecting electrode might be constructed.

33. What ions would *interfere positively* with the electrode that you described in problem 32?

Chapter 7

Qualitative Analysis of Anions

INTRODUCTION The environmental effects of the common anions of the classical Qual scheme are not, in general, very toxic. It is, more often, the danger of the associated cation that is more important from an environmental point of view. As you know by now, this was the common thread in our trip through the Qual scheme's cations groups. From Group I we choose mercury with it powerfully dangerous organomercury form. The chapter detailing Group II analyses highlighted cadmium, a metal whose prominence in batteries and the semiconductor industry unfortunately gives it extensive industrial and home use and thereby great chance for its escape into the environment. The conundrum posed by chromium was the centerpiece of the Group III elements. The fact that this metal is both required for human health and yet also toxic at somewhat higher concentrations highlights the complexity of the interaction between life and the chemicals in the environment. The cations of the soluble group, Group V, are greatly different from the previous cations that were discussed because they are, for all practical purposes, environmentally inert as charged species. Remember that in pure, neutral, elemental form most elements, sodium and potassium included, have substantially different chemical characteristics than they do as their ionic forms—elemental sodium and potassium are so reactive that they will reduced hydrogen in water and the results can be explosive! The polyatomic cation ammonium also has a significantly different character than the neutral molecule ammonia, NH_3.

SELECTED ANIONS AND TOXICITY The differences between the final classical Qual scheme Group V members that we studied, K^+, Na^+, and NH_4^+, and the earlier cations in terms of toxicity actually helps us to introduce the analyses of a few of the anions that are normally included in qualitative analysis: acetate, $C_2H_3O_2^-$, carbonate, CO_3^{2-}, chloride, Cl^-, nitrate, NO_3^-, and sulfide, S^{2-}. These also have relatively low immediate toxicity to humans. This is probably due to their solubilities and the fact that they are so ubiquitous inside of and in contact with living systems. In fact most of these anions play important parts in our bodies as well. For instance, sulfide provides important linkages in proteins; and carbonate/bicarbonate are important anions in buffering blood pH. The additional roles as described below are substantially different from those that you have been introduced to up to this point. These ions are less important in the local region where they are introduced into the environment. Instead their significance, in the case of carbonate and sulfide, is indeed global in scope.

Carbonate and Global Warming

The global cycle of carbon involves an extremely large and important movement of the element carbon among the earth's gases (atmosphere), oceans, lakes, and rivers (hydrosphere), and soils and rocks (geosphere or pedosphere). Together these regions of our planet are collectively called the **biosphere**. The biosphere contains all of the life on the planet. Deciding where the carbon cycle begins is arbitrary (think about the chicken and the egg) so we will begin our description with plants. These organisms use carbon as the foundation of the structure of leaves, stems, and roots. Carbon is biochemically combined with hydrogen and other elements to make up the biochemicals that allow plants to grow and expand their structures to increase chances for reproduction and ability to collect food, nutrients, and water. When plants die, their carbon is partially contributed back to the atmosphere as CH_4, methane, during anaerobic decay by microorganisms, and some carbon is released as carbon dioxide again. Some of the plant's **biomass** is deposited in the soil as carbonate anion and partially degraded organic compounds that can be used by other plants. They are also buried in a form of long-term carbon storage that will ultimately be release via weathering—the attack by the atmospheric processes on exposed rocks, mountains, etc. In sum, this stored carbon can be buried in sedimentation and recycled much later or it can react with normally present dissolved chemicals in precipitation; however, ultimately it will be recycled into the atmosphere as carbon dioxide.

Carbonate anion that is dissolved in water plays a very important role in the movement of carbon to the oceans. Soluble forms of carbonate and bicarbonate (HCO_3^-) and some of the soluble organic plant and animal matter flow (with many other complex anions and cations) into streams and rivers and at length into the oceans. The carbon that arrives in the oceans is absorbed to some degree by organisms who live there to make their shells or skeletons. When these animals die part of the carbon, mostly in the form of calcium carbonate, falls to the bottom of the ocean and is buried there. The rest of the oceans carbon is released, once again, to the atmosphere as carbon dioxide. The ocean floor's "stored" carbon is ultimately recycled by the hot temperatures of the earth itself which finally releases the buried carbon as carbon dioxide that slowly bubbles up from deep underground (e.g. carbonated springs) or released violently during volcanic eruptions.

By now you might have guessed the next important part of this description of the global carbon cycle. The part that human beings play: We are changing the amount of carbon in two of the reservoirs of the global carbon budget. We are adding massive amounts of carbon dioxide to the atmosphere by the burning of carbon that is normally stored in fossil fuels like petroleum, coal, and natural gas and wood. The increase in atmospheric CO_2 levels to date has been approximately 30 % since 1850 and is projected to be an increase of 75 % by the year 2060. We are doubly altering the planetary carbon budget by 1) removing carbon storage in the tropical rain forests by deforestation and 2) accomplishing this deforestation to a great degree by cutting and then burning the downed plants, thereby immediately releasing the carbon that these organisms contain. The pace of deforestation in the tropics has been greatly increasing in the last generation.

The important part that carbon dioxide plays in the global warming of our planet is very controversial and best left to the reader to research, discuss, and decide on (soon); however, this area is clearly one of the most significant that our planets faces as the recent Environmental Summit in Rio de Janeiro, Brazil in 1992 has shown. Countries from all over the planet met to discuss limiting carbon dioxide emissions into the atmosphere.

Nitrate, Nitrite, Fertilizer, and Cancer

The increased yield of modern agricultural crops can be traced to a large degree to the increased use of nitrogen-rich fertilizers (and irrigation). This has undoubtedly led to the low price and wide variety of foods available, mainly in developed countries; however, overuse of fertilizers has also led to the increase of nitrate runoff to streams and rivers from the fertilized fields. The extensive increase in large animal feedlots has also increased the amount of nitrogen in runoff and, subsequently, nitrate concentration in our population's water supplies. Finally, cured meats, like bacon, sold in the United States are cured and preserved with nitrate and nitrite in order to extend the shelf life and increase the "quality" and availability of this product to consumers. A major threat from uncured meat is from a toxin called botulin released by an organism that grows in uncured meat. Nitrite successfully inhibits this bacterium's growth. It is probably important to note that there are other preservatives available; however they are more expensive.

The above uses have increased our exposure to nitrate and nitrite ions. So what's the worry about nitrite? At the pH in the adult stomach, nitrite anion is converted to $H_2NO_2^+$. This species reacts with other chemicals normally present in that environment to yield nitrosamines. Though nitrosamines *have* been proven to cause cancer in animals they have not been directly correlated with cancer in humans; however, studies are still being performed that compare high nitrate containing diets with low nitrate containing diets and the effects on humans.

One of the environmental concerns with increased nitrate exposure lies in our body's ability to convert (reduce) this nitrate anion, NO_3^-, to nitrite anion, NO_2^- in the stomach. The reduction is not actually performed by our bodies but is instead accomplished by symbiotic (mutually beneficial) microorganisms in our digestive tract. One of these is called *Escherichia coli (E. coli.)* Part of the complication in this relationship is that adults apparently absorb ingested nitrate quickly enough in the upper digestive tract so that little or none gets converted to nitrite. Of course, ingested nitrite from preserved food is still present and can therefore create nitrosamines. In infants, the danger of nitrate exposure is a little clearer because their less acidic stomach pH allows *E. coli* to live higher in their digestive tracts and allows for more nitrate to nitrite conversion.

Though the danger of high nitrate effects on human beings is still in doubt, the results of low food production—due to no fertilizer use—and botulism—due to microbe contaminated meats—*are* well known, and this recognition of low crop yield and toxin poisoning has put us on the food preservation road that we are on now—a road from which it is hard to exit. Ultimately, the direct danger from the increased use of nitrogen fertilizers to adults may "only" be a correlation with increased chances of cancers late in life instead of acute source of cancer earlier.

SULFUR AND TEMPERATURE

The importance of the sulfur cycle has, like the carbon cycle, only recently been clearly determined. Since the concentration of gaseous sulfur compounds in the atmosphere is so much less than, for instance, carbon dioxide, the interaction between the parts of the sulfur cycle are more difficult to discern than those of the carbon cycle. However, the role that sulfide anion, S^{2-}, plays in this global wheel of sulfur movement is pivotal. Sulfur, present almost everywhere in the environment as the most oxidized and stable form, sulfate anion, interacts with bacteria and fungi in water and in soils. These microorganisms reduce the sulfur in sulfate to sulfide and then add methyl groups. The reason that this takes place is not well understood; however, it probably occurs as part of the microorganisms metabolic processing of the food and nutrients that they encounter and their need for sulfur in necessary biochemicals in their systems. After methylation, dimethyl sulfide (DMS), among other volatile sulfides, is released into the air. This volatilization occurs because the boiling point of DMS (38º C) is so close to the normal temperatures in the biosphere (25º C). In effect, it simply evaporates from these microorganisms' environments into the air.

In the atmosphere, reduced sulfur compounds are oxidized back to sulfate by natural oxidants found in the air. Small particles containing sulfate, called sulfate aerosol, act as a place on which water droplets condense to form clouds—clouds that would normally not form in the atmosphere without the presence of the aerosol. Cooling of the earth's surface occurs because of the shielding of the earth from the sun's rays by these clouds that form. Clouds act as a sun shade for the ground underneath! It is generally well accepted that this is why the plumes of clouds were so prominent in satellite photographs of the Pacific Ocean down wind of the volcanic eruption of Mount Pinatubo in the Philippines in 1991. This was an eruption that released millions of tons of sulfur and ash into the atmosphere and subsequently temporarily cooled to some degree the entire Northern hemisphere. Another famous volcanic eruption in April 1815 (in Indonesia) caused the "year without a summer" that resulted in freezing temperatures in Europe (in the Northern hemisphere) during the following year's summer months.

The effects of volcanoes in atmospheric temperatures is, as mentioned, fairly well established; however, the relationship between organisms in the biosphere and atmospheric temperature is less well understood. In the earth's oceans, photosynthetic algae near the surface release DMS into the air above the ocean. Like sulfur dioxide from volcanoes, DMS is also oxidized to sulfate aerosol that act as cloud "formers." The more algae in the ocean, the more DMS released. The more DMS is released, the more clouds form. In this way, microorganism could be affecting the amount of cloud formation over the oceans. The role of clouds in the amount of energy that reaches the earth from the sun and therefore the earth's temperature is known to be important but still not very well understood. The formation of clouds might feedback into the temperature of the ocean environment of the algae. This, therefore, would be the biospheric regulation of the planet's temperature and, if found to be true, might help explain the fact that the planet's temperature has actually varied very little over the life of the planet, about 4.8 billion years, even though the output of energy of our sun has increased significantly, + 25 to 30 %, in that time.

QUALITATIVE ANALYSIS OF COMMON ANIONS

There are many anions that are normally detected by qualitative analysis. The list can actually stretch beyond 20 if members of the same family are detected, for instance fluoride, chloride, bromide, and iodide. We will continue the selective determination of the members of this group as we have in earlier groups. The following procedures will center on only five anions: acetate, $C_2H_3O_2^-$, carbonate, CO_3^{2-}, chloride, Cl^-, nitrate, NO_3^-, and sulfide, S^{2-}. The test for all of these except chloride involves the production of gases—some safely detectable by smell, some not. Follow the procedures exactly as written and use the lab hoods when necessary to protect you and your labmates.

As in the procedures described before, you will be directed to perform the detection of a known sample of each member before working on an unknown.

Determination of sulfide

The strong smell of hydrogen sulfide, H_2S, is the key to the determination of the sulfide ion in solution in our procedure. H_2S is present in both basic and especially acidic solutions that contain S^{2-}. Perform the following tests under the hood and use care in determining the odor of H_2S—rotten eggs. A little of this toxic gas goes a long way. Put two drops of 0.05 M sodium sulfide solution in a clean test tube. Slowly fan your hand over the test tube and toward your nose. Continue this process, slowly moving and fanning the test tube toward your nose until you smell rotten eggs. You probably won't have to acidify this known sample at all to detect H_2S. If you don't detect the smell of rotten eggs with your nose close to the test tube then, holding your test tube under the hood, add one drop of 6 M HCl. Starting again at arm's length, slowly fan your hand over the test tube and toward your nose. Continue this process, slowly moving and fanning the test tube toward your nose until you smell rotten eggs.

If you perform the unknown tests below, you will be able to detect sulfide by merely smelling your sample without adding any reagents at all, unless your unknown only contains extremely trace amounts. If, however, you must acidify and retest your sample, use one drop of 6 M HCl. (Remember, you can not use this sample now for the subsequent chloride test below). After acidifying with one drop, slowly fan your hand over the test tube and toward your nose. Continue this process, slowly moving and fanning the test tube toward your nose until you smell rotten eggs or have assured yourself that no hydrogen sulfide is present in your unknown. Carefully repeat the test with two more drops of HCl if the first step is negative. Leave a positive sulfide test tube in a test tube rack under the hood until you dispose of it as directed by you lab instructor.

Determination of chloride anion

To 1 mL of a solution of 0.05 M sodium chloride solution add one drop of 0.1 M silver nitrate solution. The immediate formation of insoluble silver chloride is a clear indication of the presence of chloride ion.

If you performed the sulfide test on your unknown and sulfide anion is present, the silver test for chloride may be inconclusive because of the formation of insoluble silver sulfide. To test for chloride in the presence of sulfide, outgas the sulfide as H_2S by acidifying the sample solution first by adding (*under the hood*) 1 mL of 3 M H_2SO_4. Put the test tube in a beaker of boiling water for two minutes under the hood to boil away H_2S and then add the silver nitrate. Again, the white precipitate of AgCl is confirmation for the presence of chloride in the absence of sulfide anion.

Confirmation for acetate

Like hydrogen sulfide, acetic acid has enough vapor pressure to have a significant presence in the headspace above an acidified solution containing acetate anion. If sulfide is not present in your sample (see above) then the presence of $C_2H_3O_2^-$ can be detected by acidification with H_2SO_4. Take 1 mL of 0.1 M sodium acetate and acidify with 3 drops of concentrated sulfuric acid. Carefully perform the wafting smell test as described for H_2S but this time you will smell vinegar instead of rotten eggs.

If sulfide is present this interference can be removed by allowing H_2S to outgas after the sulfuric acid addition. This will also remove CO_3^{2-} and NO_3^-. After H_2S is no longer detectable the presence of acetate can be confirmed by adding 4 or 5 drops of ethyl alcohol. The presence of acetate will be confirmed by the presence of the odor of ethyl acetate—the smell of nail polish (not nail polish remover). If you want a sure-fire positive test add a few drops of ethyl alcohol to a few drops of pure sample of sodium acetate and carefully perform the smell test.

When testing an unknown (with none of the other anions present), if the test for acetate is negative after acidification, heat the test tube in boiling water and periodically remove and carefully repeat the smell test until you are convinced of the presence or absence of acetate.

Carbonate anion tests

Acidified carbonates will outgas carbon dioxide just as a soft drink will bubble as the bottle or can warms up. If your sample contains no sulfide, chlorides, or acetates you will be able to detect the presence of carbonate in a solution by merely acidifying with a few drops of concentrated sulfuric acid and looking for the bubbling of a clear gas. Alternately, solutions with carbonates present will also produce insoluble barium carbonate upon the addition of a few drops of 2 M $BaCl_2$. The presence of a white precipitate that forms within a few minutes after the barium chloride addition confirms the presence of carbonate.

Take 2 mL of a saturated solution of sodium carbonate and carefully and slowly acidify with 2 drops of concentrated sulfuric acid. The bubbling that you see is the release of carbon dioxide.

If your sample has been confirmed for the presence of some of the other anions in this procedure and the bubbling and $BaCl_2$ addition tests are inconclusive, a final test for the presence of carbonate is possible. Get a test tube with a saturated solution of *freshly made* barium chloride and a clean eye dropper. In a clean test tube, use one drop of concentrated H_2SO_4 to acidify a fresh, 2 mL sample of your unknown. Draw a single drop of the saturated $BaCl_2$ solution into the tip of the eye dropper. Hold the eye dropper down into the test tube as close as possible to the test tube's solution without touching the eye dropper's tip into the acidified solution contained there *and without squeezing the dropper's bulb*. If carbon dioxide is being outgassed, barium carbonate will form in the tip of the eye dropper. The presence of carbonate in your unknown will be confirmed by the formation of a milky solution in the dropper tip.

Detection of nitrate

The test for nitrate ion is based on its reduction in the presence of acid to nitrogen monoxide (NO_3^- to NO) by iron(II). The product, $Fe(NO)^{2+}$, the nitrosyliron(II) ion, is visually detected as a brown "ring."

Put 1 mL of 0.1 M sodium nitrate in a clean test tube. Carefully acidify with 2 mL (40 drops) of concentrated sulfuric acid by adding 10 drops of concentrated H_2SO_4 at a

time. Cool the *bottom* of the test tube intermittently under flowing tap water (or with ice). Don't splash any water in the top of the tube. After all 40 drops of acid have been added and the test tube is room temperature again, add 5 drops of 0.2 M iron(II) sulfate solution to the acidified solution by letting the drops slowly roll down the inside of the test tube. Try not to disturb the acid layer and don't stir the test tube. After all five drops of $FeSO_4$ have been added, a brown ring will form between the layers in a few moments. Put this positive test aside in your test tube rack for comparison with your unknown.

Dispose of the material left over after your tests in the appropriate containers. Ask your lab instructor if you do not know where or what the correct container is.

Anion unknowns

Get a test tube containing an unknown from your instructor. Record the unknown number if this is applicable. Assume that your unknown contains all the members of this group. Starting with the sulfide test, determine the presence of all of the anions in your unknown. Repeat the known confirmation tests if you are allowed to and if the need arises because of indistinct results. Take care with the smell tests! Report to your instructor the members of this group that are present in your unknown.

INSTRUMENTAL ANALYSIS OF ANIONS

The qualitative tests of the common anions introduced in this chapter are actually some of the simplest that you have been introduced to in this book except for possibly the flame tests of the Group V cations. The success of the one step odor tests or one step reduction of nitrate by iron in acid shows a major strength of qualitative analysis: simplicity and speed. While quick and simple, the detection limits (smallest amount detectable) are actually not very good using these methods; however, as noted before, there is still a place for these kinds of tests in the modern laboratory when merely the detection of the presence of these species is desired.

The primary instrumental methods for the anions that have been covered in this chapter are some of those that we have covered before. All could be determined by inductively coupled plasma since this technique is essentially a pure elemental detector and can determine each of the elements involved, S, C, O, N, and Cl. Chloride and acetate can be detected via ion chromatography. Nitrate and sulfide can be mixed with a reagent that yields a colored product. This product is then determined via UV/vis spectrometry. Since the species detected are colored (absorb in the visible wavelengths) this method is called colorimetry. Carbonate is a little more difficult because of the contamination of aqueous solutions by carbon dioxide in the air but CO_3^{2-} can also be determined instrumentally. Acidification of a carbonate solution will quantitatively (used here to mean completely) yield carbon dioxide which is outgassed from the solution when it is heated. Using chromatography (gas chromatography as a matter of fact), carbon dioxide is then separated from the water that also evaporates from the heated acidic solution. Once separated, carbon dioxide is sensitively detected and the original amount of carbonate is calculated based on the amount of CO_2 found.

CHAPTER SEVEN
PROBLEMS

1. Why are the anions introduced in this chapter not, in and of themselves, as toxic as the cations introduced before?

2. Can you find in this book any biologically useful purposes for the two most toxic metals discussed, mercury, cadmium, and chromium?

3. Write down the definition of the biosphere. In which of the three general regions of the biosphere discussed does the majority of biomass exist?

* 4. A major source of methane that is released into the atmosphere comes from rice paddies. Why?

5. List three sources that vent carbon dioxide gas to the atmosphere.

6. What effect does cutting down a rain forest have on global carbon dioxide concentrations in the atmosphere? Why?

* 7. Why has the pace of Central and South American deforestation increased recently?

* 8. From an exterior source (book, magazine, or text), find the definition of the greenhouse effect and write down a definition that would help your parents to understand this idea.

9. Write down and balance an equation showing the reduction of nitrate to nitrite by hydrogen gas. One of the products is water.

10. Why should you not be upset that *Escherichia coli* lives inside your body?

11. Children are probably endangered more by their intake of nitrate cured bacon than adults. Why?

12. What is botulism?

13. Why is sulfate aerosol important in atmospheric chemistry?

14. Why was the year 1816 called "the year without a summer"?

* 15. If the volcanic eruption that figures prominently in the answer to the last question occurred in April of 1815, why was the year without a summer 1816?

16. How does cloud formation affect the temperature of the surface of the earth underneath the cloud?

17. What would the advantage be to algae in the oceans of affecting (controlling) the temperature of the atmosphere or the oceans?

18. The detection of sulfide anion is primarily a smell test. Have you smelled this smell before and why? Was Easter connected with this memory?

19. Write down and balance the reaction of sulfide anion with protons that produce hydrogen sulfide. What drives this reaction?

20. The test for chloride anion is used in the Group I Qual scheme. How? Why?

21. What is the interference that must be removed to get a clear reading in the chloride test as presented in this procedure?

22. Write and balance the reaction of a mixture of sodium chloride, sodium sulfide, and silver nitrate.

23. Write and balance the reaction of sodium acetate with sulfuric acid that allows you to confirm the presence of acetate in a solution with no interferences.

24. Write down and balance the reaction of a solution containing sodium acetate, sodium nitrate, and sodium sulfide that has H_2SO_4 added. What drives this reaction to the right?

* 25. Why would there be any acetate anion left in solution after boiling the mixture in problem 23? What data from a reference like the CRC Handbook or Merck Index can you find to support your answer? The Merck Index is easiest to use.

26. Based on the boiling point of hydrogen sulfide, would you expect this gas to be released from environments that also produce methane? Why or why not?

27. Why does a hot soda pop fizz more than a cold one?

28. Compare the solubility of solids in liquids with the solubility of gases in liquids. Describe the effects of planetary warming on the amount of carbon dioxide that is dissolved in the oceans.

29. Write the ionic reaction (only) of iron(II) cation with nitrate anion to produce iron(III) cation, water, and nitrogen monoxide. Remember that the reaction is performed in acid.

* 30. What are the oxidation states of all of the atoms in reactants and in all of the products in the reaction that you balanced in problem 29? Which atoms were reduced and which oxidized?

Appendix 1

Solubility Product Values at 25° C

Compound	Reaction	K_{sp}
Barium carbonate	$BaCO_3 \Leftrightarrow Ba^{2+} + CO_3^{2-}$	8.1×10^{-9}
Barium sulfate	$BaSO_4 \Leftrightarrow Ba^{2+} + SO_4^{2-}$	1.08×10^{-10}
Bismuth sulfide	$Bi_2S_3 \Leftrightarrow 2Bi^{3+} + 3S^{2-}$	1.0×10^{-80}
Cadmium hydroxide	$Cd(OH)_2 \Leftrightarrow Cd^{2+} + 2OH^-$	5.00×10^{-14}
Cadmium sulfide	$CdS \Leftrightarrow Cd^{2+} + S^{2-}$	3.6×10^{-29}
Calcium carbonate	$CaCO_3 \Leftrightarrow Ca^{2+} + CO_3^{2-}$	8.7×10^{-9}
Calcium sulfate	$CaSO_4 \Leftrightarrow Ca^{2+} + SO_4^{2-}$	2.45×10^{-5}
Chromium(III) hydroxide	$Cr(OH)_3 \Leftrightarrow Cr^{3+} + 3OH^-$	6.3×10^{-31}
Chromium(III) phosphate	$CrPO_4 \Leftrightarrow Cr^{3+} + PO_4^{3-}$	2.4×10^{-23}
Cobalt hydroxide	$Co(OH)_3 \Leftrightarrow Co^{3+} + 3OH^-$	2.5×10^{-43}
Cobalt(II) sulfide	$CoS \Leftrightarrow Co^{2+} + S^{2-}$	4.0×10^{-21}
Copper(II) sulfide	$CuS \Leftrightarrow Cu^{2+} + S^{2-}$	8.5×10^{-45}
Iron(II) sulfide	$FeS \Leftrightarrow Fe^{2+} + S^{2-}$	6.00×10^{-18}
Lead chloride	$PbCl_2 \Leftrightarrow Pb^{2+} + 2Cl^-$	1.70×10^{-5}
Lead iodide	$PbI_2 \Leftrightarrow Pb^{2+} + 2I^-$	1.39×10^{-8}

Compound	Reaction	K_{sp}
Lead sulfate	$PbSO_4 \Leftrightarrow Pb^{2+} + SO_4^{2-}$	1.3×10^{-8}
Lithium carbonate	$Li_2CO_3 \Leftrightarrow 2Li^+ + CO_3^{2-}$	1.7×10^{-3}
Manganese(II) sulfide	$MnS \Leftrightarrow Mn^{2+} + S^{2-}$	2.5×10^{-10}
Mercury(I) bromide	$Hg_2Br_2 \Leftrightarrow Hg_2^{2+} + 2Br^-$	1.3×10^{-21}
Mercury(I) chloride	$Hg_2Cl_2 \Leftrightarrow Hg_2^{2+} + 2Cl^-$	1.3×10^{-18}
Mercury(I) iodide	$Hg_2I_2 \Leftrightarrow Hg_2^{2+} + 2I^-$	1.2×10^{-28}
Mercury(II) sulfide	$HgS \Leftrightarrow Hg^{2+} + S^{2-}$	1.0×10^{-51}
Nickel hydroxide	$Ni(OH)_2 \Leftrightarrow Ni^{2+} + 2OH^-$	1.60×10^{-16}
Silver bromide	$AgBr \Leftrightarrow Ag^+ + Br^-$	7.7×10^{-13}
Silver carbonate	$Ag_2CO_3 \Leftrightarrow 2Ag^+ + CO_3^{2-}$	6.15×10^{-12}
Silver chloride	$AgCl \Leftrightarrow Ag^+ + Cl^-$	1.80×10^{-10}
Silver iodide	$AgI \Leftrightarrow Ag^+ + I^-$	1.5×10^{-16}
Zinc phosphate	$Zn_3(PO_4)_2 \Leftrightarrow 3Zn^{2+} + 2PO_4^{3-}$	1.0×10^{-32}
Zinc sulfide	$ZnS \Leftrightarrow Zn^{2+} + S^{2-}$	1.3×10^{-23}

Appendix 2

Significant Figures

EXACT NUMBERS AND UNCERTAINTY

The use of numbers in chemistry and other empirical sciences is often a way of expressing the results of measurements taken with instruments or equipment. Therefore except for actually counting individual objects, the numbers derived from laboratory measurements have an **uncertainty** associated with them. A number derived by counting objects is called an **exact number** because its value has no uncertainty. Unless a mistake was made, twenty-five beakers counted on a lab bench are exactly twenty-five beakers. It is important to remember that the uncertainty that will be defined below is not due to an error on the analyst's part—this uncertainty in inherent in the measuring process. An exact number or **counting number** is exact and has no uncertainty because the process of counting relatively small numbers of objects can be accomplished perfectly. The count of all of the people in the United States would not be an exact number and would indeed have some uncertainty. Next time you see the population of the U.S. reported, see how the uncertainty of that value is reported.

If you want to know the concentration of a chemical reagent, for instance, a potassium chloride solution in your laboratory, there is numerical uncertainty built into all of the equipment used to help you determine that concentration. The balance that you use to weigh out the solid KCl has uncertainty because it cannot give you the mass with the same exactness that you can get by, for instance, counting beakers. The last digit (on the right) of the balance's display is uncertain and will affect the certainty with which you can report the final concentration of the solution. Furthermore the volumetric flask in which you dilute the KCl with water to make the solution also has an uncertainty in its volume. Read the label on any volumetric flask and it will tell you the uncertainty in its measured volume, usually in a ± percentage.

The uncertainty that is being described is handled by scientists using a concept called **significant figures**. Following a set of rules that govern the mathematical procedures that are normally used in manipulating numbers: adding, subtracting, multiplying, and dividing and including decisions about rounding, we can systematically report the values that are calculated when using empirical number derived from experiments. There are additional rules for less common procedures involving math functions like logarithms that we will not cover here.

Finally, please note that these rules are not infallible. Their purpose is to help scientists to avoid reporting results with a level of exactness or certainty that is incorrect and that leads to the wrong conclusions in an experiment. Like all rules, these too can fail to accomplish their purpose and can lead to meaningless results if applied without judgment. Let it be noted that this author includes these rules as a means of trying to stem the tide of students reporting every answer that they calculate to as many digits as their calculator will display.

DETERMINING THE NUMBER OF SIGNIFICANT FIGURES

There are two very important concepts that must be mastered to be able to successfully handle the following significant figure rules. The first is decimal places and the second is significant figures.

The number of decimal places in a number is simply the number of digits to the right of the decimal. The number 1.234 has three decimal places. The number 901.10 has two decimal places, and the number 2345.1 has one decimal place. The definition of the number of significant figures as opposed to the number of decimal places is a little more complicated:

All digits that are nonzero are significant figures:

Number	Significant figures
987	3
27645	5
867.567	6
11	2

All zeros that are *between* nonzero digits are significant. These are sometimes called captive zeros:

Number	Significant figures
9087	4
90087	5
9090887	7
101	3

All zeros that are to the right of nonzero digits *in a number with a decimal place* are significant. These are called trailing zeros:

Number	Significant figures
123.00	5
123.4050	7
.345000	6
1.0	2

All zeros that are to the right of nonzero digits *in a number without a decimal place* are not significant. They are also trailing zeros but are not held as significant because, again, there is no decimal point (some authors use a different rule here):

Number	Significant figures
10	1
1600	2
1,000,000	1
1010100	5

All zeros that lie to the left of nonzero digits are *not* significant. These are called leading zeros.

Number	Significant figures
0.1234	4
0.01234	4
0.000001	1
0.345000	6

Finally, as we noted above, all numbers derived from counting mathematical constants, or exact conversion factors are exact numbers and therefore do not affect the significant figures in any calculation:

Number	Significant figures
25 beakers	can be considered infinite
6 atoms	does not affect uncertainty in calculation
1 in the $\frac{1 \text{ kilometer}}{1000 \text{ meters}}$	not considered in uncertainty decisions

SIGNIFICANT FIGURES IN MATH OPERATIONS

The four common mathematical procedures can be divided up into two subgroups that each have their own significant figure rules: Addition and subtraction versus multiplication and division. First let's decide how to handle operations that use addition and subtraction.

The result of an addition or subtraction is limited by the number with the fewest decimal places:

$$\begin{array}{r} 123.41 \\ +\ 000.234 \\ +\ \underline{270.1\ *} \\ 393.744 \end{array} \quad \text{--------> } 393.7$$

In the problem at the bottom of the last page, the number with the asterisk controls the number of decimal places in the answer. The 393.744 answer must be reported with one decimal place and since the second decimal place is less than 5 (it is a 4), the 7 remains unchanged and the two 4s are dropped. If the number had been 393.754 it would have been rounded up to 393.8. Again, some authors systematically round differently.

$$\begin{array}{r} 333.332 \text{ *} \\ -\ 200.33341 \\ \hline 132.99859 \end{array} \quad \text{--------> } 132.999$$

In the problem above, the answer is reported with three decimal places and, again, the rounding rules were applied with the result that the last reported decimal place changed from an 8 to a 9 and the remaining 5 and 9 were dropped.

In multiplication and division the results are controlled not by the fewest number of decimal places but instead by the number of significant figures:

$$\begin{array}{r} 8.56 \\ \times\ 1.1 \text{ *} \\ \hline 9.416 \end{array} \quad \text{--------> } 9.4$$

The initial result of 9.416 is adjusted to 9.4 because the number with the asterisk has only two significant figures and, applying rounding, the next digit, a 1, is less than 5. Note that the result is reported with two significant figures as the controlling factor not with one decimal place as the controlling factor. The next problem more clearly illustrates this:

$$\begin{array}{r} 8.56 \text{ *} \\ \times\ 11.11 \\ \hline 95.1016 \end{array} \quad \text{--------> } 95.1$$

Though the result in the previous problem appears ambiguous in regards to which rule is being applied, this problem makes it clear: the initial answer of 95.1016 is adjusted to 95.1 not based on decimal places (an incorrect application of the addition and subtraction rule) but on the three significant figures in the number 8.56.

A problem involving slightly more complex numbers might occur as follows:

$$\begin{array}{r} 8.0056 \\ \div\ 0.1100 \text{ *} \\ \hline 72.778182 \end{array} \quad \text{--------> } 72.78$$

This example applies almost all our rules at once including rounding. It is a division problem so the answer is reported based on the fewest number of significant figures. The number with the fewest number of significant figures is 0.1100 because its leading zero is not significant and its two trailing zeros are. This results in four significant figures in 0.1100 and therefore four in the answer. The number 8.0056 has five significant figures because the captive zeros are between nonzero digits and are therefore significant. Finally the preliminary answer of 72.778182 becomes 72.78 because the last 7 which must be the last reported digit is rounded up.

A few words about scientific (or exponential) notation and the electronic calculator. This may help you to apply these significant figure rules to your calculators where almost all chemists' math is performed. The rules that we have laid down for the four common math procedures are entirely consistent with the powerful mathematical figure system called scientific notation. When using your calculator, don't disregard the significant figure rules just because your calculator gives you 8, 10, or on some calculators 12 digit results. Although you can program almost all scientific calculators to round and report the correct number of significant figures *on any particular problem*, the calculator completely ignores the nuances of these rules as applied to different problems. It is very probably to your advantage to set your calculator to display as many digits as possible in either decimal or scientific notation and then to apply these significant figure rules on a problem by problem basis.

The powerful use of a calculator can be exemplified by its ability to mix decimal and scientific notation in a problem and then default to the appropriate notation system in reporting the answer. For instance:

$$\begin{array}{r} 100.01 \\ \times\ 6.022 \times 10^{23} \\ \hline 6.023 \times 10^{25} \end{array}$$

The result is a very large number which is impractical to display as a normal decimal number. Calculators automatically default to scientific notation and allow you to clean up the answer yourself before you write it down. Most also allow you to program in the correct number of significant figures and will then round the last digit correctly for you. My calculator showed 6.0226022×10^{25} as the answer to this problem. When "told" to report four significant figures, it responded with 6.023×10^{25}.

Another instance is when two relatively small decimal numbers are divided:

$$\begin{array}{r} 100.01 \\ \div\ 6.022 \\ \hline 16.61 \end{array}$$

Here again most calculators know how to report the answer, and this problem will elicit a result left in decimal notation as 16.607439 and correctly reported as 16.61 if the calculator is programmed correctly. Additional programming can give you 1.661×10^1 if you so desire or are asked for the scientific notation form.

The final word about calculators is that they, like all computers, are only as smart as the operator, and the student that accepts all answers, digits, exponents, and signs, without any application of significant figure, rounding, or logic rules will have a very difficult time in chemistry indeed. Whatever you do don't report the product of 6.022 times 100.001 as 602.206022000000.

Problems to help you practice significant figure rules and rounding follow and are presented in decimal and scientific notation. Assume that there are no exact numbers in any of the problems. The correct answers are given with the problem.

APPENDIX 2
PROBLEMS

1. Multiply the two numbers that are given and report the correct answer by applying the significant figure and rounding rules detailed in this appendix. Both decimal and scientific notation forms are given in the answers that follow each problem, and you should try to make your calculator express both forms to get the most out of each problem:

	Problem	Answer
a)	6.022 x 100.01	602.3 or 6.023×10^2
b)	68.001 x 1.2×10^1	820 or 8.2×10^2
c)	10100.00 x 0.00123	12.4 or 1.24×10^1
d)	1.0001×10^{-4} x 100	0.01 or 1×10^{-2}
e)	35.000 x 1,000,500	35018000 or 3.5018×10^7

2. Divide the two numbers that are given and report the correct answer by applying the significant figure and rounding rules detailed in this appendix. Both decimal and scientific notation forms are given in the answers that follow each problem, and you should try to make your calculator express both forms to get the most out of each problem:

	Problem	Answer
a)	6.022 ÷ 100.01	0.06021 or 6.021×10^{-2}
b)	68.001 ÷ 1.20×10^1	5.67 or 5.67×10^0
c)	10100.00 ÷ 0.001203	8396000 or 8.396×10^6
d)	1.0001×10^{-4} ÷ 100 x 10^2	0.00000001 or 1×10^{-8}
e)	35.010 ÷ 1,000,500. ← notice the decimal	0.000034993 or 3.4993×10^{-5}

3. Add the two numbers that are given and report the correct answer. Answers given are in decimal form only.

	Problem	Answer
a)	100.1 + 0.1234	100.2
b)	1.011 + 1000	1001
c)	6.022 + 0.1	6.1
d)	87.65 + 1.001	87.7
e)	1 + 0.01 + 0.01 + 0.01	1

4. Subtract the two numbers that are given and report the correct answer. Answers given are in decimal form only.

	Problem	Answer
a)	100.1 − 0.1234	100.0
b)	1.011 − 1000.	−999
c)	6.022 − 1.1	4.9
d)	87.65 − 1.0010	86.65
e)	1 − 0.01 − 0.01 − 0.01	1

Appendix 3

Chemical Reagents

The reagents used in the procedures in this book are listed below 1) by chapter number listing both the specified reagents and then solutions for known (and unknown) solutions and then 2) alphabetically for all reagents used in the book. The concentrations listed are those used in the procedures; however, slightly different concentrations may be used depending on what is available in the lab and the individual Qual test. Unless otherwise specified, all dilutions are with deionized water.

REAGENTS LISTED BY CHAPTER/GROUP ANALYSIS

Chapter 2/Group I

Reagents
6 M ammonium hydroxide (add 373 mL and dilute to 1 L)
6 M hydrochloric acid (add 495 mL TO WATER and dilute to 1 L)
0.1 M mercury(I) nitrate (dissolve 28 g $HgNO_3 \cdot H_2O$ in minimal amount of HNO_3 and dilute to 1 L with H_2O)
12 M nitric acid (add 706 mL TO WATER and dilute to 1 L)
0.3 M potassium chromate (dissolve 58.2 g and dilute to 1 L)

Knowns
0.1 M lead(II) nitrate (dissolve 33.1 g and dilute to 1 L)
0.1 M mercury(I) nitrate (dissolve 28 g $HgNO_3 \cdot H_2O$ in minimal amount of HNO_3 and dilute to 1 L)
0.1 M silver nitrate (dissolve 16.98 g and dilute to 1 L)

Chapter 3/Group II

Reagents
aluminum wire 24 or 26 gauge
6 M ammonium hydroxide (add 373 mL and dilute to 1 L)
10 % by wt. ammonium sulfide solution (dissolve 100 g and dilute to 1 L)
0.1 M cadmium chloride (dissolve 22.8 g $CdCl_2 \cdot 5/2\ H_2O$ and dilute to 1 L)
3 M hydrochloric acid (add 248 ml TO WATER and dilute to 1 L)

6 M hydrochloric acid (add 495 ml TO WATER and dilute to 1 L)
0.1 M mercury(II) chloride (dissolve 27.2 g and dilute to 1 L)
6 M nitric acid (add 353 mL TO WATER and dilute to 1 L)
sodium hydrosulfite (solid)
1 M sodium hydroxide (dissolve 58.4 g and dilute slowly to 1 L)
5 M sodium hydroxide (dissolve 292 g and dilute slowly to 1 L with cooling)
1 M thioacetamide solution in water (dissolve 75.1 g and dilute to 1 L)
0.1 M tin(II) chloride (dissolve 22.6 g $SnCl_2 \cdot 2H_2O$ and dilute to 1 L)

Knowns
0.1 M bismuth(III) nitrate (dissolve 48.5 g $Bi(NO_3)_3 \cdot 5H_2O$ and dilute to 1 L)
0.1 M cadmium(II) chloride (dissolve 22.8 g $CdCl_2 \cdot 5/2\ H_2O$ and dilute to 1 L)
0.1 M copper(II) chloride (dissolve 13.4 g $CuCl \cdot xH_2O$ and dilute to 1 L or dissolve 17.0 g $CuCl_2 \cdot 2H_2O$ and dilute to 1 L)
0.1 M tin(IV) chloride (dissolve 22.6 g $SnCl_2 \cdot 2H_2O$ and dilute to 1 L)

Chapter 4/Group III

Reagents
aqua regia (VERY CAREFULLY add 250 mL of concentrated HNO_3 to 750 mL of concentrated HCl)
6 M ammonia (add 373 mL NH_4OH and dilute to 1 L)
2 M ammonium chloride (dissolve 107 g and dilute to 1 L)
1 % dimethylgyloxime (in alcohol) (dissolve 11.6 g and dilute to 1 L with (95 %) C_2H_5OH)
3 M hydrochloric acid (add 248 mL TO WATER and dilute to 1 L)
6 M hydrochloric acid (add 495 mL TO WATER and dilute to 1 L)
3 % hydrogen peroxide (purchase and store cold or dilute 100 mL 30% H_2O_2 to 1 L)
2 M nitric acid (add 38.7 mL TO WATER and dilute to 1 L)
0.1 M potassium chromate (dissolve 19.4 g and dilute to 1 L)
0.3 M potassium thiocyanate (dissolve 29.2 g and dilute to 1 L)
8 M sodium hydroxide (dissolve 467 g in water carefully and dilute to 1 L with cooling)
2 M sulfuric acid (add 111 mL To WATER and dilute to 1 L with cooling)
1 M thioacetamide solution (dissolve 75.1 g and dilute to 1 L)

Knowns
0.1 M cobalt(II) chloride (dissolve 23.8 g $CoCl_2 \cdot 6H_2O$ and dilute to 1 L)
0.1 M iron(III) nitrate (dissolve 40.4 g $Fe(NO_3)_3 \cdot 9H_2O$ and dilute to 1 L)
0.1 M nickel(II) chloride (dissolve 23.7 g $NiCl_2 \cdot 6H_2O$ and dilute to 1 L)
0.1 M potassium chromate (dissolve 19.4 g and dilute to 1 L)

Chapter 5/Group IV

Reagents
6 M ammonium hydroxide (add 373 mL NH$_4$OH and dilute to 1 L)
0.1 M ammonium oxalate (dissolve 14.2 g (NH$_4$)$_2$C$_2$O$_4$ • H$_2$O and dilute to 1 L)
barium sulfate (solid)
0.5 M calcium chloride (dissolve 114.2 g CaCl$_2$ • 5/2 H$_2$O and dilute to 1 L)
6 M hydrochloric acid (add 495 ml TO WATER and dilute to 1 L)
0.5 M magnesium chloride (dissolve 47.6 g MgCl$_2$ and dilute to 1 L)
Magneson or S. and O. Reagent (4-(4-nitrophenylazo)resorcinol
    (Aldrich catalog # 11,466-9; dissolve 0.05 g in 100 mL of 0.25 M NaOH solution)
sodium phosphate monobasic solution (saturated)
1 M sodium hydroxide (dissolve 58.4 g and dilute slowly to 1 L)

Knowns
0.1 M barium chloride (dissolve 24.4 g BaCl$_2$ • 2H$_2$O and dilute to 1 L)
0.1 M calcium chloride (dissolve 22.8 g CaCl$_2$ • 5/2 H$_2$O and dilute to 1 L)
0.1 M magnesium chloride (dissolve 9.5 g MgCl$_2$ and dilute to 1 L)

Chapter 6/Group V

Reagents
1 M ammonium chloride (dissolve 53.5 g and dilute to 1 L)
1 M potassium chloride (dissolve 74.56 g and dilute to 1 L)
1 M sodium chloride (dissolve 58.4 g and dissolve to 1 L)
6 M sodium hydroxide (dissolve 350 g in water carefully and dilute to 1 L)

Knowns
1 M ammonium chloride (dissolve 53.5 g and dilute to 1 L)
1 M potassium chloride (dissolve 74.56 g and dilute to 1 L
1 M sodium chloride (dissolve 58.4 g and dissolve to 1 L)

Chapter 7/Common anions

Reagents
2 M barium chloride (dissolve 488 g BaCl$_2$ • 2H$_2$O and dilute to 1 L)
ethyl alcohol (purchase 95 %)
6 M hydrochloric acid (add 495 ml TO WATER and dilute to 1 L)
0.2 M iron(II) sulfate (dissolve 55.6 g Fe$_3$(SO$_4$)$_2$ • 7H$_2$O and dilute to 1 L)
0.1 M silver nitrate (dissolve 17.0 g and dilute to 1 L)
0.1 M sodium acetate (dissolve 8.2 g and dilute to 1 L)

sodium carbonate solution (saturated)
0.5 M sodium chloride (dissolve 29.2 g and dissolve to 1 L)
0.1 M sodium nitrate (dissolve 8.5 g and dissolve to 1 L)
0.5 M sodium sulfide (dissolve 120.1 g $Na_2S \cdot 9H_2O$ and dilute to 1 L)
3 M sulfuric acid (add 176 mL To WATER and dilute to 1 L)
sulfuric acid (concentrated)

Knowns
0.1 M sodium acetate (dissolve 8.2 g and dilute to 1 L)
0.1 M sodium carbonate (dissolve 10.6 g and dissolve to 1 L)
0.1 M sodium nitrate (dissolve 8.5 g and dissolve to 1 L)
0.1 M sodium sulfide (dissolve 24.0 g $Na_2S \cdot 9H_2O$ and dilute to 1 L)

REAGENTS LISTED ALPHABETICALLY

aluminum wire 24 or 26 gauge
ammonia 6 M (add 373 mL NH_4OH and dilute to 1L)
ammonium chloride 1 M (dissolve 53.5 g and dilute to 1 L)
ammonium chloride 2 M (dissolve 107 g and dilute to 1 L)
ammonium hydroxide 6 M (add 373 mL and dilute to 1 L)
ammonium hydroxide 6 M (add 373 mL NH_4OH and dilute to 1 L)
ammonium oxalate 0.1 M (dissolve 14.2 g $(NH_4)_2C_2O_4 \cdot H_2O$ and dilute to 1 L)
ammonium sulfide solution 10 % by wt. (dissolve 100 g and dilute to 1 L)
aqua regia (VERY CAREFULLY add 250 mL of concentrated HNO_3 to 750 mL of
    concentrated HCl)
barium chloride 0.1 M (dissolve 24.4 g $BaCl_2 \cdot 2H_2O$ and dilute to 1 L)
barium chloride 2 M (dissolve 488 g $BaCl_2 \cdot 2H_2O$ and dilute to 1 L)
barium sulfate (solid)
bismuth(III) nitrate 0.1 M (dissolve 48.5 g $Bi(NO_3)_3 \cdot 5H_2O$ and dilute to 1 L)
cadmium(II) chloride 0.1 M (dissolve 22.8 g $CdCl_2 \cdot 5/2\ H_2O$ and dilute to 1 L)
calcium chloride 0.1 M (dissolve 22.8 g $CaCl_2 \cdot 5/2\ H_2O$ and dilute to 1 L)
calcium chloride 0.5 M (dissolve 114.2 g $CaCl_2 \cdot 5/2\ H_2O$ and dilute to 1 L)
cobalt(II) chloride 0.1 M (dissolve 23.8 g $CoCl_2 \cdot 6H_2O$ and dilute to 1 L)
copper(II) chloride 0.1 M (dissolve 13.4 g CuCl xH_2O and dilute to 1 L or dissolve
    17.0 g $CuCl_2 \cdot 2H_2O$ and dilute to 1 L)
dimethylgyloxime 1 % (in alcohol) (dissolve 11.6 g and dilute to 1 L with
    (95 %) C_2H_5OH)
ethyl alcohol (purchase 95 %)
hydrochloric acid 3 M (add 248 ml TO WATER and dilute to 1 L)
hydrochloric acid 6 M (add 495 mL TO WATER and dilute to 1 L)
hydrochloric acid 6 M (add 495 ml TO WATER and dilute to 1 L)
hydrogen peroxide 3 % (purchase and store cold or dilute 100 mL 30% H_2O_2 to 1 L)
iron(II) sulfate 0.2 M (dissolve 55.6 g $Fe_3(SO_4)_2 \cdot 7H_2O$ and dilute to 1 L)
iron(III) nitrate 0.1 M (dissolve 40.4 g $Fe(NO_3)_3 \cdot 9H_2O$ and dilute to 1 L)

lead(II) nitrate 0.1 M (dissolve 33.1 g and dilute to 1 L)
magnesium chloride 0.1 M (dissolve 9.5 g $MgCl_2$ and dilute to 1 L)
magnesium chloride 0.5 M (dissolve 47.6 g $MgCl_2$ and dilute to 1 L)
Magneson or S. and O. Reagent (4-(4-nitrophenylazo)resorcinol
    (Aldrich catalog # 11,466-9; dissolve 0.05 g in 100 mL of 0.25 M NaOH solution)
mercury(I) nitrate 0.1 M (dissolve 28 g $HgNO_3 \cdot H_2O$ in minimal amount of HNO_3 ;
    dilute to 1 L with H_2O)
mercury(II) 0.1 M chloride (dissolve 27.2 g and dilute to 1 L)
nickel(II) chloride 0.1 M(dissolve 23.7 g $NiCl_2 \cdot 6H_2O$ and dilute to 1 L)
nitric acid 2 M (add 38.7 mL TO WATER and dilute to 1 L)
nitric acid 6 M (add 353 mL TO WATER and dilute to 1 L)
nitric acid 12 M (add 706 mL TO WATER and dilute to 1 L)
potassium chloride 1 M (dissolve 74.56 g and dilute to 1 L)
potassium chromate 0.1 M (dissolve 19.4 g and dilute to 1 L)
potassium chromate 0.3 M (dissolve 58.2 g and dilute to 1 L)
potassium thiocyanate 0.3 M (dissolve 29.2 g and dilute to 1 L)
silver nitrate 0.1 M (dissolve 16.98 g and dilute to one liter)
sodium acetate 0.1 M(dissolve 8.2 g and dilute to 1 L)
sodium carbonate 0.1 M (dissolve 10.6 g and dissolve to 1 L)
sodium carbonate solution (saturated)
sodium chloride 0.5 M (dissolve 29.2 g and dissolve to 1 L)
sodium chloride 1 M (dissolve 58.4 g and dissolve to 1 L)
sodium hydrosulfite (solid)
sodium hydroxide 1 M (dissolve 58.4 g and dilute slowly to 1 L)
sodium hydroxide 5 M (dissolve 292 g and dilute slowly to 1 L with cooling)
sodium hydroxide 6 M (dissolve 350 g in water carefully and dilute to 1 L)
sodium hydroxide 8 M (dissolve 467 g in water carefully and dilute to 1 L)
sodium nitrate 0.1 M (dissolve 8.5 g and dissolve to 1 L)
sodium phosphate monobasic solution (saturated)
sodium sulfide 0.1 M (dissolve 24.0 g $Na_2S \cdot 9H_2O$ and dilute to 1 L)
sodium sulfide 0.5 M (dissolve 120.1 g $Na_2S \cdot 9H_2O$ and dilute to 1 L)
sulfuric acid (concentrated)
sulfuric acid 2 M (add 111 mL To WATER and dilute to 1 L)
sulfuric acid 3 M (add 176 mL To WATER and dilute to 1 L)
thioacetamide solution 1 M (dissolve 75.1 g and dilute to 1 L)
tin(II) chloride 0.1 M (dissolve 22.6 g $SnCl_2 \cdot 2H_2O$ and dilute to 1 L)
tin(IV) chloride 0.1 M (dissolve 22.6 g $SnCl_2 \cdot 2H2O$ and dilute to 1 L)

Appendix 4
Toxic Waste Disposal

The "safe disposal" of the toxic wastes generated by the procedures in this book is not as oxymoronic a term as it might seem. The very nature of the compounds under study means that the materials collected at the end of each procedure will be considered hazardous waste. They must be handled with care and disposed of in an environmentally safe and legal manner; however, there are a number of steps that can be taken to precipitate the dissolved ions from solutions and subsequently decrease the volume of material that must be handled. This is an environmentally conscious act in and of itself and simultaneously reduces the cost that your school must spend to maintain a laboratory that produces environmentally minded chemists.

Since toxic materials of this nature are regulated in different ways in different places the precipitation, volume reduction, and storage techniques that follow will help the students and instructors that use this book to minimize the danger to themselves and the environment. If a single laboratory repeats these procedures from semester to semester and follows these procedures, storing the resulting solids year after year, the volume of toxic materials that accrues can be extensively minimized. This means that the collection of this waste by a regulated, licensed waste hauler (the ultimate fate of all this material) need only occur once or twice in a period of many years depending on the volume generated, waste regulations, and space available for storage. *It must be reiterated that the ultimate fate of these materials must be their delivery into the possession of a licensed waste hauler.* It is your responsibility to follow the laws that apply to your waste generation, handling, storage, and disposal.

The wastes generated by the procedures presented in this book are of a known identity; and as such, they can be precipitated following relatively simple procedures. However, if different compounds are substituted due to chemical availability, alternate procedures might be necessary to assure that all of the toxic ions are precipitated.

Note: All of these steps should be carried out under a hood!

Precipitation of Group I wastes

The Group I cations are precipitated as chlorides in acid solution. To this end the solutions collected from the Group I experiments (in Chapter 2) can be precipitated by the addition of 50 % vol/vol hydrochloric acid until no more solid forms. This will precipitate Hg_2Cl_2, AgCl, and to a large degree $PbCl_2$. After precipitation ends a small amount of lead may remain in solution. This cation can also be brought down by its precipitation as $PbSO_4$ (about 1000 times less soluble than the chloride). Add small amounts of a saturated solution of technical grade sodium sulfate to the solution acidified above until the supernatant remains clear. Stir the solution for one hour then let it sit unstirred for a few days. Add more HCl and Na_2SO_4 solution the next day to confirm that precipitation is complete; if it's not, repeat the precipitation steps.

The chromate anion used to confirm lead in the Group I experiments will be tied up as $PbCrO_4$ and will remain in the solid chloride waste precipitated.

The next step is to put the beaker containing the precipitated wastes from the Group I experiments under a continually working fume hood and let as much of the aqueous supernatant evaporate as you have time. Decrease the volume of the liquid in the waste as much as possible and then transfer the waste to a thick-walled glass container (like an empty HCl reagent bottle). Screw the top on tightly! Put a list of the ions contained in the bottle on a label on the bottle and on an index card affixed to the bottles handle via string or wire. Store the bottle under a hood if space is available.

At the end of each semester repeat these precipitation and evaporation procedures and transfer the minimized waste into the labelled storage bottle. When the level in the storage bottle gets to within two inches of the top stop adding waste. Cap the bottle securely. Write the date on the labels.

Precipitation of Group II wastes

The insolubility of the sulfides are used to capture any soluble ions left over from the experiments in Chapter 3, the Group II wastes. Adjust the supernatant above any precipitates to approximately pH 7 with either sodium hydroxide or HCl. Keeping track of the volume $(NH_4)_2S$ used, slowly add a *fresh* solution of ammonium sulfide (often sold as a 20 % wt. solution) until precipitation ends then add 15 % (by volume) more ammonium sulfide. Stir the solution under the hood for one hour and then let stand over night. Add about 5 % more ammonium sulfide the next day with stirring. The traces of mercury that may have been used in the tin confirmation test will be trapped as a solid in the precipitate matrix

The next step is to put the beaker containing the precipitated wastes from the Group II experiments under a continually working fume hood and let as much of the aqueous supernatant evaporate as you have time. Decrease the volume of the liquid in the waste as much as possible and then transfer the waste to a thick-walled glass container. Screw the top on tightly! Put a list of the ions contained in the bottle on a label on the bottle and on an index card affixed to the bottles handle via string or wire. Store the bottle under a hood if space is available.

At the end of each semester repeat these precipitation and evaporation procedures and transfer the minimized waste into the labelled storage bottle. When the level in the storage bottle gets to within two inches of the top stop adding waste. Write the date on the labels.

Precipitation of Group III wastes

Though the Group III members can be precipitated with sulfide in a manner exactly like the Group II wastes, they should be kept separate from the Group II wastes, precipitated, minimized, and placed in a container of their own with their own label. The iron/cobalt thiocyanate and nickel dimethylglyoxime complexes are presumedly stably trapped in the sulfide matrix.

The next step is to put the beaker containing the precipitated wastes from the Group III experiments under a continually working fume hood and let as much of the aqueous supernatant evaporate as you have time. Decrease the volume of the liquid in the waste as much as possible and then transfer the waste to a thick-walled glass container. Screw the top on tightly! Put a list of the ions contained in the bottle on a label on the bottle and on an index card affixed to the bottles handle via string or wire. Store the bottle under a hood if space is available.

At the end of each semester repeat these precipitation and evaporation procedures and transfer the minimized waste into the labelled storage bottle. When the level in the storage bottle gets to within two inches of the top stop adding waste. Write the date on the labels.

Precipitation of Group IV wastes

The Group IV elements that must be precipitated include barium, magnesium, calcium, and possibly strontium. All will either be precipitated by the addition of a saturated solution of technical grade sodium sulfate or will precipitate as the supernatant is evaporated under the fume hood.

The phosphate and oxalate anions and the S. and O. reagent are very minor components of this solid mix. For your information the S. and O. reagent is 4-(4-nitrophenylazo)resorcinol and is labelled by a vendor that sells it in 1993 as "an irritant."

The next step is to put the beaker containing the precipitated wastes from the Group IV experiments under a continually working fume hood and let as much of the aqueous supernatant evaporate as you have time. Decrease the volume of the liquid in the waste as much as possible and then transfer the waste to a thick-walled glass container. Screw the top on tightly! Put a list of the ions contained in the bottle on a label on the bottle and on an index card affixed to the bottles handle via string or wire. Store the bottle under a hood if space is available.

At the end of each semester repeat these precipitation and evaporation procedures and transfer the minimized waste into the labelled storage bottle. When the level in the storage bottle gets to within two inches of the top stop adding waste. Write the date on the labels.

Precipitation of Group V wastes

The wastes from this group of experiments can be completely evaporated to dryness under a hood using a hot plate. Check the pH before evaporation, and the sodium hydroxide that may be present can be neutralized (to litmus) with the addition of dilute HCl. After completely evaporating to dryness the Group V waste solution and only the Group V waste solution can be scrapped the solid into the trash can. It is harmless.

Common anion wastes

The wastes generated by the chapter investigating common anions can be added to any of the *solutions* generated by the experiments in Groups II or III before they are treated as described above. The contained acetates, carbonates, chlorides, and sulfides will help to precipitate the cations generated in those procedures. For any particular semester, add the wastes from the common anion experiments to the Group II or III wastes *before* beginning the procedures described in those steps above and then follow the directions for that section.

Finally it must be said once more that there are really no satisfactory means of recycling or purifying the wastes generated by these procedures. The "bottom line" is that it is prohibitively expensive to treat these materials in any way other then to simply decrease the volume to an absolute minimum. The final low volume, highly toxic solid cakes of these compounds must ultimately be placed in a landfill. *It must be reiterated that the ultimate fate of these materials must be their delivery into the possession of a licensed waste hauler.* It is your responsibility to follow the laws that apply to your waste generation, handling , storage, and disposal.

Glossary

A

Absorption spectrum An xy plot of wavelength (x-axis) versus amount of light taken in by a substance (y-axis).
Acid rain Atmospheric precipitation with a pH substantially below that of normal rain which is pH 5.6.
Amalgam A mixture of metals like gold or silver with mercury.
Analyte The substance or chemical species of interest in an analysis.
Anion An ion that is negatively charged.
Anodizing The process of electrically coating a metal with a protective surface.
Atomic Absorption Spectrometry An instrumental method of chemical analysis that uses the light absorbing ability of atoms as a detection tool.

B

Bioconcentration The process of increasing the amount of a substance, usually a pollutant, by it's concentration (building up) in a living organism. Often this involves the ingestion of the substance from the environment and it's deposition in the tissues of the organism in higher and higher amounts over a period of time.
Biomass The total amount of living material.
Biosphere The parts of a planet that are inhabited by life.
Boiling point This is the temperature at which a liquid evaporates at normal atmospheric pressure. Technically this is the temperature at which the liquid's vapor pressure equals the pressure above the liquid.

C

Calibration A process whereby an instrument or analytical method is quantitatively standardized.
Calibration curve An xy plot of the response (output) of an instrument (y-axis) versus the amount of a standard introduced into the instrument (x-axis).
Catalyst A substance that speeds up a reaction without being itself used up in the reaction.
Cation An ion that is positively charged.
Chelation The chemical binding of a substance to another chemical. For toxic metal chelation, this is usually accomplished by a reaction of the metal with a large organic molecule or molecules that bind the metal into an even larger yet substantially less toxic complex.
Chromate Eczema An inflammation of the skin caused by contact with cement containing chromium.
Chromatogram The output of a chromatographic process.
Chromatography The science of separation.
Cuvette A small sample holding device used in spectroscopy, often made of high purity glass.

D

Decantation The physical process of separating a solid from a liquid usually by carefully pouring the liquid into another container and leaving the solid behind.
Diffraction Grating An optical element with a finely machined surface that has been prepared in such a way that white light striking it is separated into individual wavelengths (colors) as it reflect off the surface.
Dimer Two identical molecules or ions joined together into a new molecule. Mercury(I) forms a dimer (Hg_2^{2+}) in aqueous solution.
Dipole The nonsymmetrical distribution of electron density in a molecule due to its structure.
DNA Deoxyribonucleic acid—the large helical molecule that contains the information of life's heredity stored as specific chemical sequences.

E

Electronic Shells The arrangement of (negative) electrons around the positive nucleus. Traditionally these orbits are referred to as shells.

Electroplating The process of using electricity to cause a chemical reaction to occur on the surface of a piece of metal in order to coat or plate the piece with another material.

Electrostatic Attraction This is the attraction that occurs between electrically charged particles of opposite charge.

Equilibrium The point where the forward rate of reaction and reverse rates are the same. At equilibrium the amounts of the reactants and products remain the same.

F

False Negative The result of an analytical test in which the test does <u>not</u> detect the substance tested for when that substance is actually present.

False Positive The result of an analytical test in which the test <u>does</u> detect the substance tested for when that substance is not actually present.

Flow Chart In qualitative analysis, this is a graphical representation of the tests necessary to determine the presence of a number of different elements that may be present in a sample. The arrangement of the parts are such that the flow from one part to the next represents the logical order in which the test should be approached to successfully separate the chemical components that might be present.

Functional Group A set of connected atoms that takes part in a reaction. In ion chromatography, the functional group determines the binding or chromatographic characteristics of the polymer containing the attached functional group.

G–L

Hollow Cathode Lamp The light source used in atomic absorption spectroscopy. Each lamp contains a metal surface that give off (emits) light specific to that element when the lamp is turned on.

Homeostasis The maintenance of the balance of ions and molecules in the body's organs, tissues, and blood.

Hyperammonemia A condition in which there is an elevated amount of ammonium ions in the blood.

Immiscible Two liquids are immiscible when they will not mix.

Inductively coupled plasma A very hot phase of matter in which most molecules are completely broken into the atoms that make them up. Many of the atoms present are partially stripped of their electrons. The resulting mixture of electrons and cations absorbs energy from a radio frequency source located nearby. This energy absorption further heats the plasma.

Instrumental Analysis Method of chemical detection based on nonclassical means. Usually these newer techniques involve electronic equipment that measure matter's physical characteristics such as the absorption of emission spectrum, conductivity, or mass to charge ratio.

Ion An atom or group of atoms that is electrically charged.

Laser A radiation source that produces light with special characteristics. The light generated by a laser is monochromatic (one color), high intensity (amplified), and finely focussed.

Light Absorption The process of taking in light instead of allowing it to pass through or be reflected. Light that is not transmitted or reflected is absorbed.

M–O

Matrix The chemicals in a sample that are not the analyte or substance of interest.

Molarity (M) Strictly defined in chemistry as the number of moles of solute per liter of solution.

Molar Solubility The amount in moles of a compound that will dissolve in one liter of solution at a specific temperature.

Monochromator An electronic instrument that separates light into individual wavelengths.

Nebulizer In atomic absorption spectroscopy or inductively coupled plasma, this is the sampling part of either of those instruments that disperses a liquid sample into a flowing gas stream as very fine droplets. This aerosol is then fed into the flame or plasma.

NIMBY Syndrome An acronym that represents the words Not In My Back Yard. This is a selfish, yet normal, human opposition of people to the idea of locating toxic sources (waste dumps, etc.) near them even though these same people do want the toxic material to be collected and stored somewhere.

Ore Tailings The solid material or waste left over after the mining process to separate the more valuable metal.

Osmosis The diffusion of ions through a membrane.

P

Particulates Individual particles, small enough to remain suspended in air or smoke.

pH The negative log of hydrogen ion concentration, [H^+].

Photomultiplier Tube An electronic device that detects light by electronically increasing the signal created by the arrival of photons.

Photons Individual packets of light or electromagnetic radiation.

Phototube A simple light detector that works by creating a flow of electrons in response to the arrival of light.

Plasma A very hot, highly ionized gas made up of free electrons, ions and neutral particles.

Polar Possessing two opposite attributes at once.

Polarity In molecules, polarity is caused by the unequal distribution of electron density or charge.

Pollution The contamination of the environment by the discharge of harmful materials.

Precipitation In quantitative analysis this is the production of a solid or precipitate in a solution. In the atmosphere this is the condensation of water to form rain or snow.

Q–R

Qualitative Analysis A procedure that determines the *identity* of chemicals present in a sample.

Quantitative Analysis A procedure that determines the *amount* of a particular chemical present in a sample.

Quantized Limited to discrete levels. Electrons in atoms and molecules are limited to specific sets of energy states. Electronic orbitals are quantized.

Radicals Reactive molecules whose instability is based on the lack of electron pairing.

Reference A starting point. In chemistry this is a material of known composition with which other materials can be compared.

Reference Beam In a spectroscopic instrument this is a path along which rays of light pass. The reference beam usually does not pass through the sample being analyzed.

Retention Time A measure of the time it takes a compound to move through a chromatographic column.

S

Saturated In qualitative analysis, a solution is saturated with a particular chemical species when no more of that species will dissolve in the solution at that temperature.

Semiconductor A solid material that has electrical conductivity less than a pure conductor (like copper) but more than an insulator (like glass).

Signal Transducer A device whose job it is to convert an instrumental signal from one form to another. A photomultiplier is an example of a signal transducer.

Solubility A measure of the relative amount of a solute that will dissolve into a solvent.

Solubility Product Constant A quantitative measurement of the amount of a slightly soluble salt that will dissolve in a liquid. It is derived from the mathematical product of the concentrations (in molarity) of the dissolved ions, each raised to the stoichiometry number accompanying that ion in the balanced equation.

Solute The chemical dissolved in a solvent. The chemical component present in lesser amount that the solvent.

Solvent In a mixture, it is the substance in which the solute or the component of smaller amount is dissolved. The solvent generally does not react chemically.

Spectroscopy The science of the study of spectra.

Spectroscopist An individual who studies the optical spectra of matter.
Standard Solution A chemical solution whose concentration is known with a high degree of certainty.
Stoichiometry The relationship between the amounts of the reactants and products of a reaction.
Supernatant Also supernate. The liquid floating above a precipitate. The supernatant is the liquid collected by pouring off the solution above the solid.

T–Z

Trace Metal A metallic element normally present in small concentrations in the environment. Cadmium and mercury are trace metals.
Transmission Spectrum An xy plot of light wavelengths (x-axis) versus amount of light passing through a substance (y-axis).
Wet Chemical Methods Traditional analytical means of determining chemical composition. These usually involve solutions, separations, distillations, or gravimetric analysis. Though it is not always the case, instrumental methods are often more sensitive that wet chemical methods for a specific chemical species.

Index

A

absorption spectrum 53
acid rain 80
adenine triphosphate 62
amalgam 12, 28
Amazon River 12
ammonium cycle 79, 80
anaerobic decay 96
analyte 36, 54
anion analysis 99
anions 2, 95
anodizing 27
arthritis 12
atmosphere 1, 96
atomic absorption spectrometry 18, 21, 34, 36
atomic spectra 49
ATP 62
axon 81, 87

B

barium 61
batteries 28
best fit line 38
bioconcentrated 11
biomass 96
biosphere 96
boiling point 13
botulin 97
botulism 97

C

cadmium 27
calibration curve 36, 38
carbon monoxide 28
catalyst 11
cell potential 87
central nervous system 81
chelation 29
chemical reactions 1
chromate eczema 44
chromatogram 70
chromatography 67
chromium 43

cigarette smoking 28
co-ions 69
colorimetry 102
column elution 69
common anion analysis 99
complexes 1
corrosive icon 7
counter ion 68
counting number 107
cube 2
cuvette 52

D

decantation 14
deoxyribonucleic acid 62
diffraction grating 35, 36
dimer 13
dimethyl sulfide 98
diode lase 27
dipole 1
dipole moment 1
DNA 45, 62

E

E. coli. 97
electrical potential 87
electrode 87
electrode potential 87
electromotive force 87
electronic shells 18
electrons 18, 27, 34
electroplating 27
electrostatic attraction 2
electrothermal vaporizer 36
elemental mercury 15
EMF 87
emission 82
endoscope 61
equilibrium 3, 4
Escherichia coli 97
exact number 107

F

false negative 17
false positive 17
FDA 17
fire extinguisher icon 7
flame emission 82
flame test 31
flow chart 15, 16
Food and Drug Administration 17
formaldehyde 28

G

Gaia hypothesis 98
gas chromatography 67, 102
gel chromatography 67
geosphere 96
global warming 96
gold 12
graphite furnace 19, 36
Group I analysis 13
Group II analysis 30
Group III analysis 45
Group IV analysis 63
Group V analysis 84

H

hallow cathode lamp 19, 20
homeostasis 62, 81
hydrogen ion concentration 87
hydrolysis 62
hydrosphere 96
hyperammonemia 79

I–K

icons 7
immiscible 2
inductively coupled plasma 34, 39
instrumental analysis 37
instrumental chemical analysis 17
Instrumental methods of analysis 17
ion exchange chromatography 67, 73
ion selective electrode 87, 88, 90
ions 2
itai-itai disease 28

L

lab safety 6
laptop computers 28
laser 27
leukemia 12
light absorption 17

light emission 82
like dissolves like 2
liquid chromatography 67
loading 69

M

Madame Curie 63
magneson 65
matrices 34
matrix 29, 55
measuring electrode 88
membrane electrode 88
mercuric chloride 12
mercury 11, 27
methane 96
methyl mercury chloride 12
methylation 98
microorganisms 97
Minamata Disease 11, 28
minerals 43
mobile phase 68
molar solubility 4
molarity 3
molecular spectra 49
Mount Pinatubo 98
multiple sclerosis 12

N

nebulizer 36
nerve cells 81
nerve impulse 82
neuron 62
NiCd batteries 28
nickel-cadmium 28
NIMBY syndrome 30
nitrogen 1
nitrosamines 97
Nobel Prize 89
nonpolar 2
nuclear magnetic resonance 61
nucleus 18, 27

O

ore tailings 28
organic mercury 11
organomercury 11
osmosis 62
oxygen 1

P

paper chromatography 67
pedosphere 96

pH 80, 87
pH electrode 87, 88
photomultiplier 20, 35
photomultiplier tube 20, 35
photons 20
phototube 20
plasma 34
poison icon 7
polar 1, 2
polarity 2
pollution 27
portable computers 28
potassium emission 82
precipitation 5
protons 18

Q

qualitative 17
qualitative analysis 1, 19, 30, 34, 45, 63, 84, 99
qualitative solubility rules 5
quantitative 71
quantitative analysis 19, 36
quantized 18

R

radicals 45
radioactive polonium 28
radium 63
reference beam 54
reference electrode 87
regenerating 68
resting potential 82
retention time 72
ribonucleic acid 62
RNA 62
Rowland circle 36

S

safe disposal icon 7
safety 6
safety glasses icon 7
safety icons 7
saturated 3
semiconductor 27
signal transducer 20
significant figures 107
sodium emission 82
solid state electrode 88
solubility 2, 4
solubility product 3, 4, 5
solubility product (table) 105
solubility product constant 4

solubility rules 5
solute 2
solvation 3
solvent 1
spectroscopists 54
standards 20, 37
stoichiometry 3
subshells 18
sulfate aerosol 98
suppressor column 71
surface weathering 96
symbiosis 97

T

thioacetamide 30
torch 34
trace metal 27
transmission spectrum 53

U–Z

ultraviolet/visible (UV/vis) spectroscopy 51, 56
uncertainty 107
urea 79
vitamins 43
voltmeter 87
water 1
weathering 96
wet chemical methods 17
zinc 27